"Efforts to obstruct climate action are a major factor in delaying meaningful reductions in carbon emissions. This new book captures the latest peer reviewed literature and weaves an entertaining, easy to read and chilling narrative about how a number of concentrated efforts by vested economic and ideological interests have successfully worked to imperil the planet. Read this book carefully when you develop new measures to advance climate action."

Robert J. Brulle, *Brown University, USA*

"*Climate Obstruction: How Denial, Delay and Inaction are Heating the Planet* is an authorive introduction laying out the key definitions and components of climate obstruction. To a lay reader and an audience familiar with the topic they encapsulate a clear summary of what obstruction is usefully using a three-part typology to identify and help understand the complexity of climate obstruction and why it manifests. A fantastic text and well-needed read to help understand the history of and continuing role that climate obstruction plays in delaying the required changes to mitigate the climate crisis."

Ruth E. McKie, *De Montfort University, UK*

Climate Obstruction

In *Climate Obstruction: How Denial, Delay and Inaction are Heating the Planet*, Kristoffer Ekberg, Bernhard Forchtner, Martin Hultman and Kirsti Jylhä bring together crucial insights from environmental history, sociology, media and communication studies and psychology to help us understand why we are failing to take necessary measures to avert the unfolding climate crisis.

They do so by examining the variety of ways in which meaningful climate action has been obstructed. This ranges from denial of scientific evidence for human-induced climate change and its policy consequences to (seemingly sincere) acknowledgement of the scientific evidence while nevertheless delaying meaningful climate action. The authors also consider all those actions by which often well-meaning individuals and collectives (unintendedly) hamper climate action. In doing so, this book maps out arguments and strategies that have been used to counter environmental protection and regulation since the 1960s by, first and foremost, corporations supported by conservative actors but also far-right ones as well as ordinary citizens.

This timely and accessible book provides tools and lessons to understand, identify and call out such arguments and strategies, and points to actions and systemic and cultural changes needed to avert or at least mitigate the climate crisis.

Kristoffer Ekberg is a researcher at Chalmers University of Technology. His work focuses on the political history of climate change and the environment, corporate anti-environmentalism as well as social movements and utopian thought.

Bernhard Forchtner is Associate Professor at the School of Media, Communication and Sociology and member of the Institute for Environmental Futures, University of Leicester, UK. He works on the far right and environmental communication.

Martin Hultman is Associate Professor in science, technology and environmental studies at Chalmers University of Technology, Sweden. He leads Center for Studies of Climate Change Denial (CEFORCED), as well as research groups analyzing Gender & energy and Ecopreneurship.

Kirsti M. Jylhä is a researcher at the Institute for Future Studies, Stockholm, Sweden. Her work focuses on psychological obstacles and drivers of climate engagement (e.g., climate change beliefs and emotions) as well as sociopolitical attitudes and ideologies.

Climate Obstruction
How Denial, Delay and Inaction are Heating the Planet

Kristoffer Ekberg, Bernhard Forchtner, Martin Hultman and Kirsti M. Jylhä

LONDON AND NEW YORK

Cover image: © Shutterstock

First published 2023
by Routledge
4 Park Square, Milton Park, Abingdon, Oxon OX14 4RN

and by Routledge
605 Third Avenue, New York, NY 10158

*Routledge is an imprint of the Taylor & Francis Group, an
informa business*

© 2023 Kristoffer Ekberg, Bernhard Forchtner, Martin
Hultman and Kirsti M. Jylhä

The right of Kristoffer Ekberg, Bernhard Forchtner, Martin
Hultman and Kirsti M. Jylhä to be identified as authors of
this work has been asserted in accordance with sections 77
and 78 of the Copyright, Designs and Patents Act 1988.

British Library Cataloguing-in-Publication Data
A catalogue record for this book is available from the British
Library

ISBN: 978-1-032-01925-3 (hbk)
ISBN: 978-1-032-01947-5 (pbk)
ISBN: 978-1-003-18113-2 (ebk)

DOI: 10.4324/9781003181132

Typeset in Times New Roman
by SPi Technologies India Pvt Ltd (Straive)

Contents

Acknowledgements

While writing these final lines of *Climate Obstruction: How Denial, Delay and Inaction are Heating the Planet*, we not only look back at a difficult period characterised by lockdowns, home-schooling, summer droughts and illness but also of fulfilling collaboration as we laboured through the various stages of a manuscript so important to all of us.

Indeed, although the topic was seldom uplifting, we benefited from and enjoyed the opportunity this project afforded. As such, we want to thank Routledge for enabling us to work together, coming from different disciplines and knowledge backgrounds. Collaboration was most intense as we set the scene and synthesised our findings in Introduction and Conclusion, respectively, but was at least as intriguing in Chapters 2, 3, 4 and 5 (which were initially drafted by Ekberg, Hultman, Forchtner and Jylhä, respectively). We hope that the outcome of this work will speak not just (not primarily!) to academia but to interested individuals, social movements, churches, labour unions and others.

That said, we want to extend our gratitude and appreciation to all those who helped us with this project. Formas, the Swedish Energy Agency, the Swedish Research Council, the division of Science, Technology and Society, the Centre for Studies of Climate Change Denial, and the Institute for Futures Studies for funding and institutional support. The international network of CSSN has provided fertile intellectual ground. We furthermore thank David Larsson Heidenblad, Björn Lundberg, Balša Lubarda and Niklas Olsen for providing crucial feedback to individual chapters, as well as Kjell Vowles and Magnus Linton, who have extensively commented on the entire manuscript. First and foremost, however, we want to thank Lise Benoist who, for several months, was a central part of this book project in her role as Chalmers research assistant. Lise not only dealt with images, typos, endnotes and the review of key literature but became an integral part of the collaborative effort. We want to thus thank her for her essential contribution; moreover, as Lise takes up a PhD position at Uppsala University, we also want to wish her all the best as her research career fully unfolds. Most importantly, however, we want to thank our loved ones for their support in this endeavour into the short-comings of humanity and hope that this book can make a tiny contribution to their and other (non-)humans' quality of life on this planet.

1 Introduction

Welcome to a heating planet?

It's getting hotter, isn't it? Yes, indeed, hardly a year passes without new temperature records, worsening droughts and raging wildfires across the globe. Accordingly, and as we, the authors, started to draft this book, the opening of COP26 in October 2021, the international climate summit held in Glasgow, Scotland, was accompanied by the United Nations *Emissions Gap Report 2021*, which warned that existing plans to limit the climate catastrophe set us on a path towards 2.7°C higher temperatures.[1] Since then, we have learned that the pledges made at COP26 could limit warming to just below 2°C.[2] The latter is undoubtedly good news, though these pledges do not only need to be realised in the first place but would still leave us far away from what was identified as preferable in Paris in 2015, that is, to limit overall warming to 1.5°C. All this happened against ever more visible consequences of our (in)actions, as in the case of the Australian bushfires in 2019–2020 (Figure 1.1). Indeed, as we finalise this introduction in mid-2022, populations around the world are experiencing the effects of climate change, from heatwaves to floods and rising sea levels, and the just-released Assessment Report by Working Group II of the Intergovernmental Panel on Climate Change (IPCC) urgently reminds us that

> Climate change is a threat to human well-being and planetary health. Any further delay in concerted anticipatory global action on adaption and mitigation will miss a brief and rapidly closing window of opportunity to secure a liveable and sustainable future for all.[3]

But still, "[w]e are sleepwalking to climate catastrophe", as the Secretary-General of the United Nations António Guterres said in

DOI: 10.4324/9781003181132-1

Figure 1.1 Bush fire devastation in Australia. The 2019–2020 bushfires in Australia reportedly killed 1.25 billion non-human animals and 33 humans and burnt up to 19 million hectares.[4] Research has linked a trend of increasing burned area and forest megafires to climate change.[5] Despite coal's disastrous CO_2 record, the main driver of human-induced climate change, Australia, has remained one of the world's largest coal exporters.

Source: Adwo / Alamy Stock Photo.

March 2022. While the consequences of this historic failure will only be truly visible in decades to come, the outcome does raise the question of how we – and this is primarily the "we" of those historically responsible for greenhouse gas emissions – have diverged from the responsibility to curb emissions and end fossil fuel combustion? Or, to draw on Guterres' metaphor, why haven't we woken up? This book is an attempt to answer the question: why have we, from major companies to ordinary citizens around the world (though, of course, with very different levels of responsibility and capacities to act), not taken appropriate measures to avert the unfolding climate crisis which is itself part of a much broader ecological crisis?

Granted, the early 2020s have been challenging in additional ways, with COVID-19 rightly dominating global news and politics alike. Indeed, the pandemic quickly overshadowed the *Fridays for Future* youth protests and other climate movements, preventing further activism. By now, the latter has bounced back, but so have greenhouse gas emissions which, due to various restrictions intended to bend the pandemic, from factory activity to travel restrictions, had declined.[6] In such a context, COP26 was supposed to mark the official return of climate policy, possibly being the final opportunity to set standards and implement goals to avoid hardship and suffering of all living beings on our warmer planet. The fact that such distress is already upon us – from East to West, from North to South – should have further strengthened the efforts by all those involved. In fact, responses to COVID-19 have shown that societies around the world are able to act rather decisively in the face of an imminent crisis, including both changing behaviours and unleashing vast financial resources. So, even though there are some significant differences, such as time scale and proximity of the threat, could this pandemic help us see how coordinated, large-scale action on climate change might happen too?[7] Maybe it is not too late to limit the further heating of the planet.

Yes, there are encouraging signs, including the aforementioned, global youth protest (Figure 1.2) and some changes in social norms in parts of the world regarding high-emitting behaviours, for example, eating meat and flying. Indeed, studies on public perception further evidence change. The 2021 Special Eurobarometer showed that 93 per cent of Europeans considered climate change to be a serious problem, with 64 per cent having already taken at least one action that tackles climate change over the past six months. Awareness and media coverage seem

Figure 1.2 "Youth 4 Climate" march in London, UK, on February 15, 2019.

Source: Guy Corbishley / Alamy Stock Photo.

to increase also in the USA, a crucial country due to its large share of global greenhouse gas emissions and particularly powerful obstruction.[8] As to global trends, results indicate that climate concern is increasing in many countries, even though these trends need to be carefully followed as climate perceptions are likely to fluctuate in response to crucial events.[9] For example, time will tell if events such as the COVID-19 pandemic and the Russian war against Ukraine will have effects on the public perceptions of the climate crisis.

What is, however, known already is the link between rising greenhouse gas emissions – in particular carbon dioxide (CO_2) – and rising temperatures, and it was already in the 19th century that what is nowadays known as the greenhouse effect was first identified. That is, heat (which enters as sunlight) is trapped in the atmosphere by greenhouse gases such as CO_2 and methane. And as humans have increasingly caused such emissions, global temperatures have risen too (Figure 1.3). In fact, if these emissions are not curbed, projections point to catastrophic rises in temperatures. And even if all current pledges are implemented, the average temperature is going to rise by almost 2°C, which will have severe impacts on lives on Earth.

It is against this background that we want to offer an introduction to the various dimensions of obstructing necessary climate change mitigation. We explore these dimensions by bringing together diverse backgrounds and fields of expertise: communication studies and sociology, environmental history and psychology. By offering such a

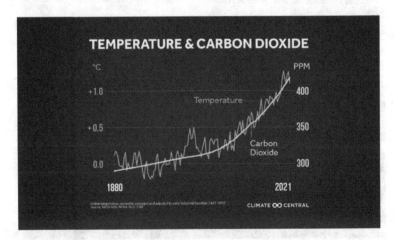

Figure 1.3 140 years of rising CO_2 and rising temperatures (Climate Central 2022).[10]

synthesis of disparate fields of research, we hope to offer a new perspective, something which starts with our choice of terminology as we, in line with a more recent trend, speak of climate obstruction[11] (instead of widely used terms such as denial and scepticism). We will outline the reasons for this in greater detail below, but for now, let us simply clarify that obstruction serves as an umbrella term for various obstacles which stand in the path of effective climate mitigation. These range from literal denial of anthropogenic climate change to the opposition, delay or dismissal of effective climate policies, at corporate, governmental, societal and individual levels, for various economic, political and psychological reasons. Indeed, the term obstruction strikes us as being particularly useful at this juncture, as being able to bring together a range of conceptualisations from different academic fields and societal grassroot organisations to understand why adequate actions have not taken place yet.

Moreover, obstruction intriguingly points to the question of who obstructs, of agency. While everyday actions reproduce an unsustainable system, not everyone contributes and benefits equally from it, or has equal powers to change it. To exemplify, the fossil fuels lobby does not obstruct in the same way as the individual consumer does. All too often today, an emphasis is put on our individual consumption and carbon footprint by, for example, companies and governments.[12] In fact, it was British Petroleum, as part of its obstructionist public relations operation, that introduced a carbon footprint calculator already in 2004. What we argue is that (unintentional) everyday contributions to obstruction are always embedded and must be viewed in relation to environmental (counter-)movements, the economy, science and policy. In other words, while individual choices, for example, to fly less or turn vegetarian/vegan, can be beneficial, they are clearly not enough for the change needed. Rather, a wider approach is required, one based on an acknowledgement of the embeddedness of these actions in a web of socio-economic relationships.[13]

In this context, *Climate Obstruction: How Denial, Delay and Inaction are Heating the Planet* introduces different actors with different agendas – though all of them are implicated in obstruction. We start broadly and consider the ways in which the climate has been understood over the past decades and centuries, but quickly focus on those companies – fossil fuels ones in particular – whose interests have distorted public debate. This "denial machine" has long been in full swing, engaging various strategies and, despite knowledge from fossil fuel industries' own research and development departments, put profit over people. And yet, party-political actors too have failed, at times including those at the centre or the political left who are today, usually, most likely to support climate action. Speaking of politics further-more points to parties on the (far) right which, over the past one to two decades, have commonly been obstructionist – an obstruction informed not simply by a need to maximise profits but by ideological maxims. And then, there are we, ordinary citizens of the Global North who, for decades, have enjoyed an over-consumerist lifestyle.[14] Sure, there are substantive (carbon) inequalities within the Global North too, but many smoking chimneys have been relocated to the Global South, leaving us, the (relatively) privileged, consuming at unsustainable levels.[15] Figure 1.4 illustrates this responsibility.

This has happened in the context of global environmental governance which emerged during the late 1960s, with the issue of climate change becoming a question for national politics during the early 1970s and, subsequently, international politics in the mid-1980s. During the early 1990s, the conflict between environment and industry was actively reframed in terms of self-regulation, voluntary codes of conduct, technological fixes, marketisation of emissions and environmental declarations. That is, the possibility to decouple environmental damage from economic growth, often by means of more efficient technologies, was talked about in terms of, for example, "sustainable development" or "ecological modernisation".[17] In the mid-1980s, this framing was constructed as a top-down, multilateral project under the auspices of the United Nations in the form of United Nations Framework Convention on Climate Change (UNFCCC) and IPCC, both born out of a wish for scientific agreement and political compromise.[18] However, this arrangement hinges on consensus in the cold war context, and while it has thus enabled participation from many countries, it has also led to scientific reports and statements that are overly cautious in not shaking up the status quo.[19] Further, the ideal of compromise has left (non-)state drivers of climate change – think of oil, coal and gas companies – unmentioned and has thus restricted systemic changes.[20] In short, the IPCC-driven approach to global climate policy has to be seen in light of an unequal world and the ambition to produce a single, authoritative account which all countries could follow.

However, since the failed negotiations in Copenhagen in 2009, researchers have argued that the regime governing environmental and climate change issues has changed into a more polycentric form, in which diverse national governments play a decisive role. Importantly, this has also changed the way opposition to climate mitigation policy

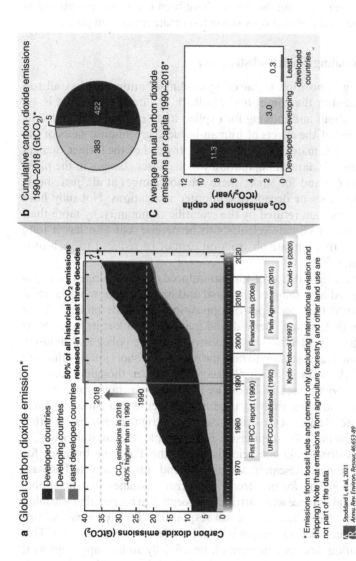

a Global carbon dioxide emission*

Developed countries
Developing countries
Least developed countries

50% of all historical CO₂ emissions released in the past three decades

CO₂ emissions in 2018 ~60% higher than in 1990

First IPCC report (1990)
UNFCCC established (1992)
Kyoto Protocol (1997)
Financial crisis (2008)
Paris Agreement (2015)
Covid-19 (2020)

Carbon dioxide emissions (GtCO₂)

Stoddard I, et al, 2021
Annu. Rev. Environ. Resour. 46:653-89

b Cumulative carbon dioxide emissions 1990–2018 (GtCO₂)*

383   422

c Average annual carbon dioxide emissions per capita 1990–2018*

CO₂ emissions per capita (tCO₂/year)

Developed 11.3
Developing 3.0
Least developed countries 0.3

* Emissions from fossil fuels and cement only (excluding international aviation and shipping). Note that emissions from agriculture, forestry, and other land use are not part of the data

Figure 1.4 Global carbon dioxide emissions, when and where.[16]

Source: reproduced from Stoddard *et al.* 2021, CC BY.

is created and enacted.[21] In particular, with the Paris agreement and voluntary targets being set at the level of the state, opposing the authoritative account of IPCC has become less meaningful while controversies surrounding the climate have been connected to nationalistic politics and populist opposition to climate mitigation policy.

Understanding climate obstruction

Analysing opposition to meaningful climate mitigation has all too often ended in the conceptual deadlock of "denialism", that is, of thinking about and looking for explicit and "extreme" cases of outright denial of the effects of human-induced greenhouse gases or the existence of climate change altogether. Instead of the former, deniers claim that it is natural fluctuation which causes changes in the planet's climate – and for some, there are no changes at all, just short-term anomalies or deceptive computer simulations. Not only have these claims been refuted by the scientific community, by more than 99 per cent of climate scientists to be precise, but also has it long been clear that climate obstruction cannot be reduced to such outright denial.[22]

While outright denial of human-induced climate change is increasingly viewed as a thing of the past, and hence attention has increasingly turned to other types of opposition to climate mitigation, we illustrate that, in fact, literal denial quite early on lost its function as being the only effective way to obstruct climate policy and industry regulation. Here, we want to take the reader on a short journey through some of the ways in which obstruction has been thought about and used in previous research before we introduce our stance further.

Already early on, analyses tried to make sense of a particular form of climate change obstruction, namely outright denial or, as often used synonymously, scepticism. For example, Stefan Rahmstorf, a leading German climate scientist from the Potsdam Institute for Climate Impact Research, identified trend, attribution and impact scepticism.[23] The former, trend scepticism, assumes that no global warming takes place while attribution scepticism accepts that warming exists, but claims humans are not responsible. Finally, impact scepticism assumes that warming might be harmless or even beneficial. This classification has been influential, beneficially so in many ways as it contains a useful separation on denialist foci. In fact, this conceptualisation of denial is still drawn upon in many studies – though in recent years, it has been integrated into much more comprehensive frameworks to understand the complexities of climate obstruction. As such, outright denial is analysed as a part of a wider set of types of knowledges and practices detrimental to climate mitigation, something which subsequently became captured in, for example, the distinction between "epistemic" and "response scepticism".[24] While the former captures the three types introduced by Rahmstorf and is related to beliefs on the status and existence of climate change (science), the latter concerns doubts related to the actions (policy responses) undertaken to tackle climate change. For example, the energy transformation picking up (in varying speeds) across the world has caused opposition by pointing to those who are said to (rightly or wrongly) not being able to carry the costs.

Before we continue with considering such typologies, let us however take note of another aspect: Rahmstorf and other researchers speak of types of *scepticism*. We do not argue against using this term but note possible objections in the context of dismissal of scientific claims. Since the birth of Western science, a sceptic attitude has carried positive connotations, especially in science itself. Scepticism, understood in this way, is key to scientific endeavour (which relies on exchanging and scrutinising claims within the scientific community, a process which can lead to a de facto consensus indicating trustworthy knowledge).[25] Indeed, in an age of fake news, a sceptic attitude open to rational debate is even key for democracies. Thus, are the various forms of obstruction, starting with what Rahmstorf introduced as trend, attribution and impact scepticism, "sceptic" in the first place? For example, one of the authors of this book initially used the concept of scepticism (since that is how they called themselves) but changed terminology to denial when he turned to empirical analyses of internal, party-political policy documents, lobbyism and digital media.[26] However, not only have many of those writing about climate change obstruction used the term scepticism, some have even made the point that "denial" carries unhelpful connotations due to its association with Holocaust denial.[27] While this might sound far-fetched, some of those not accepting the scientific consensus on human-induced climate change have in fact used this trope strategically to reject being labelled "denialist" – though others have, through their own communication, illustrated that climate and Holocaust denial are at times going hand in hand.[28]

Whatever position is taken on this terminological issue, the division between science and policy response has paved the way for many more typologies which, in various ways, have added further nuance. Maybe the most detailed one separates core scepticism concerning scientific evidence, what was above called epistemic scepticism, from concomitant forms of scepticism, that is, process (scepticism towards the scientific processes behind the evidence for anthropogenic climate change/the way public debates around climate change) and response scepticism (which we have already encountered).[29] Possibly even more helpful is the extensive list of specific arguments which has been attached to this threefold way of understanding climate obstruction, details which illuminate how evidence, process and response scepticisms are realised. While people do not need to be evidence sceptics when being process or response sceptics, these three should be understood as a spectrum. This, in turn, highlights that the sceptic logic "flows" over these categories, making too sharp distinctions unhelpful. Indeed, while response scepticism is furthest away from evidence scepticism, these obstructionist arguments can be used simultaneously. Similar attempts to capture both outright denial and other modes of obstruction are provided by Lorraine Whitmarsh, as well as Häkkinen and Akrami, whose multi-item scales range from (not very common) rejections of human-made climate change to expressions of some degree of uncertainty and doubt, including claims that the media is often too alarmist about the issue.[30] Yet another attempt to cover this entire spectrum includes the use of neutralisation techniques, including excusing a continued use of fossil fuels, as varying responses to the threat of climate change.[31] These are "denial of responsibility", "denial of injury", "denial of victim", "condemnation of the condemner" and "appeal to higher loyalties".[32]

Another classification scheme used in climate research, the one proposed by Stanley Cohen, suggests three ways in which denial occurs in society.[33] First, literal denial simply involves a claim that something is not happening – such as that the Earth is not experiencing higher global average temperatures. Second, interpretive denial in and through which actors reinterpret what is happening in a way that it loses significance. One example of such interpretive denial would be to view global warming as positive for humanity and more-than-human nature. The third form of denial is implicatory denial. Here, Cohen focuses on moral, political and psychological implications and captures ways in which individuals, institutions and states refuse to act (or obstruct possible actions) despite knowing they should. What differentiates Cohen's from many other classifications is his interest in *who* denies. In line with this, we acknowledge the importance of actors in our analysis of obstruction and, furthermore, view different types of obstruction as potentially appearing simultaneously and overlapping.

Literally or interpretatively denying the reality of human-made climate change was always only one way to obstruct climate policies. Indeed, academic analysis has increasingly looked at those who accept climate change but nevertheless chose not to act (adequately). This covers the aforementioned process and response types of scepticism, but also, for example, a proposal by William Lamb and his colleagues, in 2020.[34] The latter speak of four discourses of delay, separating the redirection of responsibility, from pushing for non-transformative solutions, emphasising the downsides of climate mitigation and simple surrender as we cannot – allegedly – do anything anyway.

What we take away from this literature is that any attempt to understand climate obstruction today must be able to not just capture explicit denial. Rather, it must be able to offer a framework integrating a much wider field of obstructionist strategies. Indeed, terms like denial and scepticism are increasingly unsatisfying in their ability (or lack of!) to describe how we fail in addressing the climate crisis. For example, climate denial risks depicting a far too simplistic, reductionist dichotomy between climate "deniers" and "non-deniers", which conceals the complexity of climate inaction. After all, "non-deniers" might nevertheless bear responsibility for opposing, delaying or undermining adequate climate policies. Scepticism, in turn, has been criticised for being too weak a term and for, as we discussed above, possibly even granting an aura of scientific legitimacy to those not acting. In response, there has been a push recently to define obstruction in terms of action "to delay policies that seek to reduce greenhouse gas emissions or general pollution by seeing them as a threat to business".[35] Others have similarly described fossil-based corporations as the main obstructors, speaking of "a regime of obstruction" which, interestingly, unfolds at different scales, from everyday life in which we, as consumers, reproduce fossil fuel dependencies to transnational policies, which maintain our reliance on these fuels.[36]

To offer a coherent, interdisciplinary terminology that allows detailed analyses, we will speak of climate change obstruction to understand the complex ways in which the status quo is reproduced, be it in the dimension of science, politics and culture, or the economy. More specifically, and to capture the complex nature of such relations of obstruction, we differentiate the latter into primary, secondary and tertiary obstruction (Figure 1.5).

First, primary obstruction includes all those wilful or ignorant activities which have come to be known under the labels of denialism and epistemic/evidence scepticism. The paradigmatic examples of such activities are the doings of the denial machine and the claims of politicians which have aimed to undermine action to mitigate climate change by questioning the scientific consensus concerning human-made climate change. For example, such primary obstruction includes

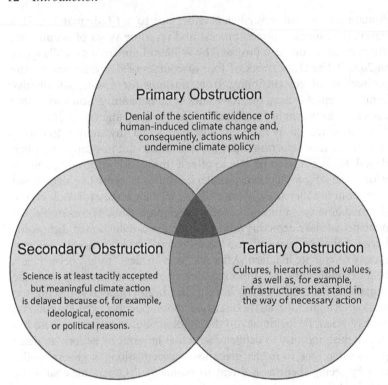

Figure 1.5 Three types of obstruction. While different, they can and do overlap.

claims that not human-induced emissions but "the sun" is responsible, which subsequently justifies attacks on policies set to mitigate climate change. Second, secondary obstruction includes all those calls which do not deny the human-induced nature of the climate crisis (science), but nevertheless delay or forestall meaningful climate action. This happens, for example, by emphasising that "one should take a rational, non-alarmist approach" and that "one should move cautiously as we cannot destroy our industry, while others do nothing". Finally, tertiary obstruction denotes all those processes by which often well-meaning individuals and collectives live their lives in denial, for example, through various coping mechanisms. Our (in)actions might not intend to obstruct climate action, but the (unintended) consequences of these actions do hamper climate action. In other words, even though the science is accepted, and one might act in (some kind of) accordance with this acceptance, be it when consuming or voting, business remains "as usual".

While these three types of obstruction need to be separated, they need to be equally acknowledged as they all contribute towards the same outcome: lack of taking urgently needed steps. Indeed, and as we indicate in Figure 1.5, these three types of obstruction are not simply connected, but overlapping and interwoven in manifold, complex ways, cutting across the political, economic and wider cultural spheres.

While we highlight actors and their doings, our proposal does not ignore wider material conditions and power relations. Yes, structures, such as capitalist relations, patriarchy (translating into certain gender roles and masculinities) as well as nationalist and petrocultural identities, are key factors behind doing nothing against climate change. But we stress that these structures need actors who invent, enforce and uphold them. Indeed, not only outright attempts to deny are done and need doers, but ignorance and silence too, although often thought of in terms of absence, of not doing, are brought into being.[37]

Of course, when it comes to responses to the climate crisis, not every disagreement is obstructionist. For example, such disagreement might simply be because the proposal does not go far enough to meaningfully prevent the unfolding of the climate crisis. That is, governments might be criticised for their (lack of) action, and those who put forward such criticism will in fact make or have made more far-reaching suggestions. Similarly, proposals might be criticised for overly burdening the already disadvantaged (within the Global North and/or between North and South); here too, if alternatives are suggested, such criticism cannot count as obstruction. Finally, we, the authors of these lines, accept that climate change is caused by human emissions – though we, simultaneously, concede that changes we have made to our lifestyles are not going far enough. This is an issue certainly not limited to the four of us, but it is also not captured by what we have come to know above as "response scepticism". Indeed, such terminological aspects will be further addressed in subsequent chapters, but they point to disciplinary particularities which affect our understanding of climate change obstruction. This book is an effort to learn across and make use of such disciplinary knowledge bases.

What we do – and what we do not do in this book

We have organised this book around four key chapters which follow this introduction – before offering concluding remarks and possible lessons from almost half-a-century of climate obstruction.

We begin, in Chapter 2, with a description of the role of fossil fuels in society and the way climate science has developed since the first discussions on CO_2 and atmospheric change in the mid-19th century. By tracing the diverse responses to the issue of environmental destruction as a global problem up until the late 1980s from industry actors and states, we show the historical construction of key trends that today are mobilised to obstruct climate and environmental action. In other words, this chapter recognises and emphasises the long history of these issues and brings together a wide range of empirical, in-depth studies on the entanglements of climate knowledge and societal change.[38]

Chapter 3 looks primarily at the denial machine and dissects the parts that made up the denialist climate change countermovement between the inauguration of the IPCC in 1988 and the Trump presidency in the U.S. It also discusses secondary obstruction as climate change was regularly framed by proponents of ecological modernisation theory, such as Green Keynesians, in terms of being a market failure. Here, we discuss crucial actors and their strategies deployed to counter meaningful action during a period when more than half of all historic greenhouse gas emissions were emitted.

Chapter 4 moves on to discuss the far right and its climate obstruction. Here party-political and non-party actors have employed a range of arguments, from conspiracy theories of climate change as a global hoax to claims that climate change policies are leading to deindustrialisation and economic despair. We introduce this class of actors before turning to distinct modes of obstruction they employ, that is, the claims they make when engaging in primary obstruction, but also, and much more so, in secondary obstruction. As we will see, such climate obstruction often overlaps with what will have been discussed in Chapters 2 and 3, though it takes a particular twist in the case of the far right.

In Chapter 5, we take a step away from the political field to look at the psychology of climate change and highlight that the denial machine and politicians have successfully sown doubt among the public. Although outright denial is relative rare, even the most solid and basic claims of climate science are still debated in society. There is also disagreement regarding the nature of the threat and the preferable solutions, as well as a persistent gap between expressed concern and engagement with the issue. In this chapter, we therefore discuss the interplay of individual opinions and actions, socio-economic relationships, information environment and societal structures. Moreover, we describe the roles of psychological factors that help explain views and behaviours that contribute to climate obstruction, including basic information processing, norms, ideological attitudes and identity processes.

Finally, our conclusion brings these various perspectives together, highlights the past, present and future conflicts around climate change and its obstruction, and indicates possible ways to move forward.

In short, *Climate Obstruction: How Denial, Delay and Inaction are Heating the Planet* considers different types of actors and domains, from the economy to the political and the everyday. In so doing, we hope to offer an accessible and comprehensive map to understand the failure to mitigate climate change. Of course, we are not the first to attempt so and thus build on similar, indeed illuminating accounts.[39] What we add is an integration of multiple perspectives and, consequently, different levels of society in doing so. And yet, we recognise that our attempt is also limited, in that our expertise is concentrated on the Global North. While climate change affects countries in the Global South disproportionally – effects which cannot be understood without also considering imperialism and colonialism – the active obstruction of climate action is, to our knowledge, a phenomenon which has primarily emanated from the Global North. In other words, the obstruction of climate action is also a continuation of the historical injustices manifest in levels of emissions and their (in)direct consequences.

However, this does not mean that actors in the Global South lack agency or are not contributing to the problem. Indeed, besides the well-publicised and funded campaigns of Western environmentalists, movements and local protest groups in the Global South have stood at the forefront, both in actions against fossil fuel extraction and with demands to the international community (Figure 1.6) – while anti-environmentalism and climate action has seemingly also proliferated in, for example, the BRIC countries. This is maybe most visible in cases of primary obstruction by Bolsonaro in Brazil but also as secondary obstruction by heads of state, state institutions and in social media in India, China, the Philippines and Trinidad and Tobago.[40] The presence of climate obstruction in the Global South begs the question of whether this looks different from the USA, where most research has been done, as transitioning away from an energy source and export revenue seems to come with similar economic and cultural problems in both the Global North and South.[41]

Building on knowledge generated by different actors and in a wide range of contexts, and by integrating expertise from communication studies and sociology, environmental history and psychology, this book is thus an attempt to offer a (not so) short history of climate obstruction. In so doing, it attempts to show how we have arrived at

Figure 1.6 Vanessa Nakate, climate activist from Uganda in Glasgow, Scotland, in 2021.

Source: Allstar Picture Library Ltd / Alamy Stock Photo.

this point and indicate that (and how) obstruction can be overcome. To do so, the next chapter starts by turning to the middle of the 20th century and considers how climate change obstruction and contestations of environmental politics emerged and evolved.

Notes

1 United Nations Environment Programme (2021) "Emission Gap Report 2021". Available at: https://www.unep.org/resources/emissions-gap-report-2021 [Last accessed April 26, 2022].

2 Meinshausen, M. *et al.* (2022) "Realization of Paris Agreement pledges may limit warming just below 2°C", *Nature*, 604, pp.304–309. https://doi.org/10.1038/s41586-022-04553-z.

3 IPCC (2022) "Summary for policymakers" in Pörtner, H.O. *et al.* (eds) *Climate Change 2022: Impacts, Adaptation, and Vulnerability. Contribution of Working Group II to the Sixth Assessment Report of the Intergovernmental Panel on Climate Change.* Cambridge: Cambridge University Press; Mitchell, A. and Chaudhury, A. (2020) "Worlding beyond 'the' 'end' of 'the world': white apocalyptic visions and BIPOC futurisms", *International Relations*, 34(3), pp.309–332. https://doi.org/10.1177/0047117820948936.

4 WWF (2020) "Australian bushfires: Recovering and restoring Australia after the devastating 2019-20 bushfires". Available at: https://www.wwf.org.au/what-we-do/bushfires [Last accessed March 29, 2022].

5 Canadell, J.G. *et al.* (2021) "Multi-decadal increase of forest burned area in Australia is linked to climate change", *Nature Communications*, 12(6921). https://doi.org/10.1038/s41467-021-27225-4.

6 Tollefson, J. (2021) "Carbon emissions rapidly rebounded following COVID pandemic dip", *Nature*, 4 November. Available at: https://www.nature.com/articles/d41586-021-03036-x [Last accessed March 29, 2022].

7 Hochachka, G. (2020) "Unearthing insights for climate change response in the midst of the COVID-19 pandemic", *Global Sustainability*, 3, E33. https://doi.org/10.1017/sus.2020.27.

8 Leiserowitz, A. *et al.* (2022) *Global Warming's Six Americas, September 2021*. Yale University and George Mason University. New Haven, CT: Yale Program on Climate Change Communication.

9 Capstick, S. *et al.* (2015) "International trends in public perceptions of climate change over the past quarter century", *WIREs Climate Change*, 6, pp.35–61. https://doi.org/10.1002/wcc.321.

10 Climate Central (2022) "Global temperature anomalies averaged and adjusted to early industrial baseline (1881-1910)". Available at: https://ccimgs-2022.s3.amazonaws.com/2022GHG/2022GHG_CO2AndTemp_en_title_lg.jpg [Last accessed May 13, 2022].

11 For prominent, though not clearly defined uses of obstruction, see McKie, R.E. (2021) "Obstruction, delay, and transnationalism: examining the online climate change counter-movement", *Energy Research & Social Science*, 80(102217). https://doi.org/10.1016/j.erss.2021.102217 and Brulle, R.J., Hall, G., Loy, L. and Shell-Smith, K. (2021) "Obstructing action: foundation funding and US climate change counter-movement organizations", *Climatic Change*, 166(17). https://doi.org/10.1007/s10584-021-03117-w.

12 Kaufman, M. (2020) "The carbon footprint sham: a 'successful, deceptive' PR campaign", *Mashable*. Available at: https://mashable.com/feature/carbon-footprint-pr-campaign-sham [Last accessed March 29, 2022].

13 Blumenfeld, J. (2022) "Climate barbarism: adopting to a wrong world", *Constellations*, pp.1–17. https://doi.org/10.1111/1467-8675.12596.

14 Steffen, W. *et al.* (2015) "The trajectory of the Anthropocene: the great acceleration", *The Anthropocene Review*, 2(1), pp.81–98. https://doi.org/10.1177/2053019614564785.

15 This trend has become slightly less clear since the mid-1990s. See Baumert, N. *et al.* (2019) "Global outsourcing of carbon emissions 1995–2009: a reassessment", *Environmental Science & Policy*, 92, pp.228–236. https://doi.org/10.1016/j.envsci.2018.10.010.

16 Stoddard, I. *et al.* (2021) "Three decades of climate mitigation: why haven't we bent the global emissions curve?", *Annual Review of Environment and Resources*, 46(1), p.656. https://doi.org/10.1146/annurev-environ-012220-011104.

17 Aronczyk, M. and Espinoza, M. (2022) *A Strategic Nature: Public Relations and the Politics of American Environmentalism*. Oxford: Oxford University Press; Bergquist, A.-K. and David, T. (2022) "Beyond limits to growth: collaboration between the international business and United Nations in shaping global environmental governance", *Les Cahiers de l'IEP, IEP Working Paper Series*, 80.

18 Howe, J.P. (2014) *Behind the Curve: Science and the Politics of Global Warming*. University of Washington Press.
19 Brysse, K., Oreskes, N., O'Reilly, J. and Oppenheimer, M. (2013) "Climate change prediction: erring on the side of least drama?", *Global Environmental Change*, 23(1), pp.327–337. https://doi.org/10.1016/j.gloenvcha.2012.10.008.
20 Carton, W. (2019) "'Fixing' climate change by mortgaging the future: negative emissions, spatiotemporal fixes, and the political economy of delay", *Antipode*, 51(3), pp.750–769. https://doi.org/10.1111/anti.12532; Aykut, S.C. and Castro, M. (2017) "The end of fossil fuels?: understanding the partial climatisation of global energy debates" in Aykut, S.C., Foyer, J. and Morena, E. (eds) *Globalising the Climate: COP21 and the Climatisation of Global Debates*. Milton, UK: Routledge, pp.173–193.
21 Lövbrand, E., Hjerpe, M. and Linnér, B.-O. (2017) "Making climate governance global: how UN climate summitry comes to matter in a complex climate regime", *Environmental Politics*, 26(4), pp.580–599. https://doi.org/10.1080/09644016.2017.1319019; Aykut, S.C. (2016) "Taking a wider view on climate governance: moving beyond the 'iceberg,' the 'elephant,' and the 'forest'", *WIREs Climate Change*, 7, pp.318–328. https://doi.org/10.1002/wcc.391.
22 A useful and accessible introduction to climate science, widely shared climate-obstructionist claims and how to respond is provided by Darling, S.B. and Sisterson, D.L. (2014) *How to Change Minds about Our Changing Climate: Let Science Do the Talking the Next Time Someone Tries to Tell You …: The Climate Isn't Changing: Global Warming Is Actually a Good Thing: Climate Change Is Natural, Not Man-made: … and Other Arguments It's Time to End for Good*. New York: The Experiment. While scientists have long agreed that climate change is happening – and that it is happening because of human influence, the number of 97 per cent in particular has been paraded. The latter was based on a survey covering climate science between 1991 and 2012, and was published in 2013. See Cook, J. *et al.* (2013) "Quantifying the consensus on anthropogenic global warming in the scientific literature", *Environmental Research Letters*, 8(024024). https://doi.org/10.1088/1748-9326/8/2/024024. A study published in 2021, however, covered almost 90.000 climate-related papers published between 2012 and 2020 found that more than 99 per cent of the peer-reviewed scientific literature accepts the thesis of anthropogenic climate change: Lynas, M., Houlton, B.Z. and Perry, S. (2021) "Greater than 99% consensus on human caused climate change in the peer-reviewed scientific literature", *Environmental Research Letters*, 16(114005). https://doi.org/10.1088/1748-9326/ac2966.
23 Rahmstorf, S. (2004) "The climate sceptics" in Munich Re (ed) *Weather Catastrophes and Climate Change: Is There Still Hope for Us?* Munich: Munich Re, pp.76–83.
24 Capstick, S.B. and Pidgeon, N.F. (2014) "What is climate change scepticism? Examination of the concept using a mixed methods study of the UK public", *Global Environmental Change*, 24, pp.389–401. https://doi.org/10.1016/j.gloenvcha.2013.08.012.
25 Oreskes, N. (2019) *Why Trust Science?* Princeton and Oxford: Princeton University Press.

26 Furthermore, see Lewandowsky, S., Mann, M.E., Brown, N.J.L. and Friedman, H. (2016) "Science and the public: debate, denial, and skepticism", *Journal of Social and Political Psychology*, 4(2), pp.537–553. https://doi.org/10.5964/jspp.v4i2.604.

27 Van Rensburg, W. (2015) "Climate change skepticism: a conceptual re-evaluation", *SAGE Open*, 5(2), pp.1–13. https://doi.org/10.1177/2158244015579723; Almiron, N. and Moreno, J. (2022) "Beyond climate change denialism: conceptual challenges in communicating climate action obstruction", *Ámbitos*, 55, pp.9–23. https://doi.org/10.12795/Ambitos.2022.i55.01.

28 On climate change conspiracy theories, see Uscinski, J., Douglas, K. and Lewandowsky, S. (2017) "Climate change conspiracy theories" in Nisbet, M. *et al.* (eds) *The Oxford Encyclopedia of Climate Change Communication*. Oxford: Oxford University Press. For an example of such a connection, see Forchtner, B. (2019) "Articulations of climate change by the Austrian far right: a discourse-historical perspective on what is 'allegedly manmade'" in Wodak, R. and Bevelander, P. (eds) *"Europe at the Cross-road": Confronting Populist, Nationalist and Global Challenges*. Lund: Nordic Academic Press, pp.159–179.

29 Van Rensburg, W. (2015) "Climate change scepticism: a conceptual re-evaluation", *SAGE Open*, pp.1–12. https://doi.org/10.1177%2F2158244015579723.

30 Häkkinen, K. and Akrami, N. (2014) "Ideology and climate change denial", *Personality and Individual Differences*, 70, pp.62–65. https://doi.org/10.1016/j.paid.2014.06.030; Whitmarsh, L. (2011) "Scepticism and uncertainty about climate change: dimensions, determinants and change over time", *Global Environmental Change*, 21, pp.690–700. https://doi.org/10.1016/j.gloenvcha.2011.01.016.

31 Sykes, G.M. and Matza, D. (1957) "Techniques of neutralization: a theory of delinquency", *American Sociological Review*, 22(6), pp.664–670.

32 McKie, R.E. (2019) "Climate change counter movement neutralization techniques: a typology to examine the climate change counter movement", *Sociological Inquiry*, 89, pp.288–316. https://doi.org/10.1111/soin.12246.

33 Cohen, S. (2001) *States of Denial: Knowing about Atrocities and Suffering*. Cambridge, UK: Polity Press.

34 Lamb, W.F. *et al.* (2020) "Discourses of climate delay", *Global Sustainability*, 3(E17), pp.1–5. https://doi.org/10.1017/sus.2020.13.

35 Almiron, N. and Moreno, J. (2022) "Beyond climate change denialism".

36 Carroll, W.K. (2021) *Regime of Obstruction: How Corporate Power Blocks Energy Democracy*. Edmonton, Alberta: AU Press, p.482.

37 See, for example, Proctor, R.N. and Schiebinger, L. (2008) *Agnotology: The Making and Unmaking of Ignorance*. Stanford University Press; Zerubavel, E. (2007) *The Elephant in the Room: Silence and Denial in Everyday Life*. Oxford: Oxford University Press.

38 Oreskes, N. and Conway, E.M. (2011) *Merchants of Doubt: How a Handful of Scientists Obscured the Truth on Issues from Tobacco Smoke to Global Warming*. Bloomsbury Publishing USA; Ekberg, K. and Hultman, M. (2021) "A question of utter importance: the early history of climate change and energy policy in Sweden 1974–1983", *Environment and History*, 8. https://doi.org/10.3197/096734021X16245313030028.

39   For example, and illustrating the different types of books available on this topic, some more, some less academic: Aronczyk, M. and Espinoza, M. (2022) *A Strategic Nature: Public Relations and the Politics of American Environmentalism*. Oxford: Oxford University Press; Klein, N. (2015) *This Changes Everything: Capitalism vs. the Climate*. London: Penguin; Oreskes, N. and Conway, E.M. (2010) *Merchants of Doubt*; Washington, H. and Cook, J. (2011) *Climate Change Denial. Heads in the Sand*. London: Earthscan; Hulme, M. (2009) *Why We Disagree about Climate Change: Understanding Controversy, Inaction and Opportunity*. Cambridge: Cambridge University Press; Tindall, D., Stoddart, M. and Dunlap, R.E. (eds) (2022) *Handbook of Anti-Environmentalism*. Cheltenham, UK: Edward Elgar Publishing.

40   Menezes, R.G. and Barbosa Jr., R. (2021) "Environmental governance under Bolsonaro: dismantling institutions, curtailing participation, delegitimising opposition", *Zeitschrift für vergleichende politikwissenschaft*, 15(2), pp.229–247. https://doi.org/10.1007/s12286-021-00491-8; Gagliardi, J., Oliveira, T., Magalhães, S. and Falcão, H. (2021) "'The Amazon is ours'. The Bolsonaro government and deforestation: narrative disputes and dissonant temporalities" in Bødker, H. and Morris, H.E. (eds) *Climate Change and Journalism: Negotiating Rifts of Time*. London: Routledge, pp.155–169; Atkins, E. and Menga, F. (2021) "Populist ecologies", *Area*, 00, pp.1–9. https://doi.org/10.1111/area.12763; CSSN (2021) "Is climate obstruction different in the global south? Observations and a preliminary research Agenda", *CSSN Briefing*, 4; Liu, J.C.E. and Zhao, B. (2017) "Who speaks for climate change in China? Evidence from Weibo", *Climatic Change*, 140(3–4), pp.413–422. https://doi.org/10.1007/s10584-016-1883-y; Tynkkynen, V.P. and Tynkkynen, N. (2018) "Climate denial revisited: (Re)contextualising Russian public discourse on climate change during Putin 2.0", *Europe-Asia Studies*, 70(7), pp.1103–1120. https://doi.org/10.1080/09668136.2018.1472218; Yagodin, D. (2021) "Policy implications of climate change denial: content analysis of Russian national and regional news media", *International Political Science Review*, 42(1), pp.64–77. https://doi.org/10.1177/0192512120971149; McDermott Huges, D. (2017) *Energy without Conscience: Oil, Climate Change and Complicity*. Durham: Duke University Press. On the Philippines, see Marquardt, J., Oliveira, M.C. and Lederer, M. (2022) "Same, same but different? How democratically elected right-wing populists shape climate change policy-making", *Environmental Politics*. https://doi.org/10.1080/09644016.2022.2053423.

41   CSSN (2021) "Is climate obstruction different in the global south?"; Norgaard, K.M. (2011) *Living in Denial: Climate Change, Emotions, and Everyday Life*. Cambridge, MA: MIT Press; McDermott Huges, D. (2017) *Energy without Conscience*.

2 The foundations of climate obstruction

Introduction

Where and when did climate change obstruction begin? The most simplified and commonly provided answer is to pinpoint it to the unusually warm summer of 1988. This year the Intergovernmental Panel on Climate Change (IPCC) was established and NASA scientist James Hansen testified before the U.S. Congress that climate change was already detectible, two events that sparked a concerted response from the industry.[1] However, the conflicts and thoughts being activated at that time – by what has come to be known as the "denial machine" further described below – were not new. The history of climate change obstruction stretches far back and literal denial is only one of the methods used. It is thus not surprising that recent studies trying to get a grip of the various forms of denial and delay have looked back in time for answers.[2] In retrospect, internal company documents as well as notes from party-political negotiations show how knowledge and intentional withholding and concealment of knowledge have been used for decades to protect the continuation of extraction, refining and burning of fossil fuels.[3]

In this chapter, we unearth the foundations of climate change obstruction and show how delaying tactics, while being more forcefully and concertedly activated from the late 1980s onwards, have origins further back in time. Beginning with describing the role of coal, oil and gas in the modern industrial society and the emergence of climate science, the chapter then moves on to the early contestations of environmental politics. From the onset of the conflict between industry and ecological science in the late 1960s, we trace how economic theories, public relation (PR) strategies and political manoeuvring have been used to keep decarbonisation and other effective climate change politics off the table up until the late 1980s, while having simultaneously pioneered some of the tactics still used in present-day delay.

DOI: 10.4324/9781003181132-2

Fossil capital and the geopolitical weight of oil

Any history of anthropogenic global heating must begin with the main culprit, the combustion of fossilised organic material. Fossil fuels ushered the world into a new era in the 19th century. With the wholesale introduction of a movable and storable energy source, the possibility to control and manage energy supply and create demand was at the same time put in place. It was precisely these capabilities of fossil fuels, and the whole socio-technical system enacted with it, that made them triumphant as energy sources.[4] Since the boom of fossil fuels, and especially oil, in the early 20th century, enormous wealth and geopolitical power has been accumulated by a small portion of the world's population. As researchers have shown, the geographical expansion of fossil fuel corporate activities in the beginning of the 20th century served the dual purpose of, first, securing energy for industry and populations in the Global North and, second, limiting extraction to keep prices high.[5] The control of oil (and coal, and later gas) was crucial for the developments of asymmetries of power on a global scale during the late 19th and early 20th centuries.[6] One such example is the way Western oil companies and the USA, through the control of oil findings in the Middle East, enforced scarcity to maintain high prices and domestic revenues.[7] As production was running at full force in the 1950s and 1960s, oil was quite literally what made the wheels speeding up into a Great Acceleration.[8] For many states, fossil fuels, and especially oil, had become a cultural signifier of industrial modernity and the lifeblood of the society in the 1950s.[9] To mention just one example of the visuals spread at the time, a Ford commercial portrayed the fossil-fuelled car as the liberation of suburban women and as an absolute necessity for family life.[10] Some 20 years later, and on the other side of the Atlantic, the chair of the energy commission in Sweden could proclaim, "Private car ownership is a human right, an absolute necessity for freedom" when confronted with the risk of climate change.[11] To be clear, limiting the access or possibility to extract fossil fuels through regulation while the risk of its emissions was recognised would mean a total transformation of the business of many companies, but would also threaten the material foundation upon which first European and then U.S. geopolitical dominance had been built.[12]

When evidence of the risks for disruptive ecological catastrophe due to burning fossil fuels began to appear in the scientific community, among fossil companies' own research departments, and among political leaders in the late 1960s, the need for regulation materialised – and so did organised opposition to it. However, before we move to the pivotal years of the late 1960s, let us quickly relate the way climate science developed during the period from the late 1800s.

Scientific knowledge and the idea of "beneficial climate change"

Already in the 19th century, experiments showed that higher concentrations of CO_2 meant a greater capability for the atmosphere to conserve incoming heat from the sun. Eunice Newton Foote (1819–1888) is the first scientist known to have experimented on the warming effect of sunlight on different gases. In her paper, *Circumstances Affecting the Heat of the Sun's Rays*, presented at the American Association for the Advancement of Science conference in 1856, she theorised that a changed proportion of carbon dioxide in the atmosphere would consequently change its temperature.[13] While such crucial observation was made more than 150 years ago, the science on the sources and effects of greenhouse gases was still underdeveloped and did not result in any attempts to limit emissions.[14] For example, when the theories of scientist Svante Arrhenius were explained to the public in Sweden in the early 20th century, the main contributor of CO_2 emissions to the atmosphere was thought to be volcanic eruptions and the effects of a tripling of CO_2 concentrations were believed to transform the world into "a complete paradise".[15] Such depictions have been picked up by obstructionists in more recent times, especially among "carbon vitalists", who argue that an increase of CO_2 concentrations will only result in a greener planet, proclaiming CO_2 as the "Gas of life".[16] However, as the burning of fossil fuels intensified at the beginning of the 20th century, the cause of climate change was increasingly acknowledged as having an anthropogenic origin. One of the first to recognise this was meteorologist Guy Stewart Callender who, already in 1938, stated that the temperatures were rising due to CO_2 emissions from the industry. The positive connotations of such a planetary warming nevertheless remained present, as Finnish chemist Kurt Buch stated in 1951: "since it is industry which has delivered it [CO_2], it has apart from other permissible work now also managed to provide some comforting heat to old Tellus".[17] In the Soviet Union, such depictions remained and were expanded even in the late 1980s, where scientists argued that climate change would produce a net benefit primarily to the agricultural sector, which was and is an important sector in the region.[18]

While the science on climate change was inconclusive in the 1950s and as several scientists argued that the result of a hotter planet would be beneficial, the perception of climate change as a risk was nonetheless present even among fossil fuel corporations at the time. The fossil fuel industry actively engaged in research focusing on the potential destructive force their product had on the Earth because they needed to know what to expect from a warmer planet and when to make their infrastructure investments in the ocean or the permafrost.[19] The physicist Edward Teller warned the fossil fuel industry of temperature rise due to CO_2 emissions already in 1959, a year before the publication of, among climate scientists, the crucial Keeling curve measuring the increase in CO_2 concentrations in the atmosphere.[20] A few years later, Frank Ikard, then president of the American Petroleum Institute, informed the organisation of the content of a new report from the U.S. president's Science Advisory committee, which stated that the combustion of fossil fuels would "cause marked changes in climate". Ikard told the audience that "The report unquestionably will fan emotions, raise fears, and bring demands for action".[21] The positive or inconclusive connotations of layering our atmosphere with carbon dioxide from burning fossil fuel were beginning to change. However, in ads from the early 1960s fossil fuel companies could still brag about the effects of their product as shown in Figure 2.1.

Another issue that shifted the framing of climate change relates to funding. In the Cold War era of the 1950s, climate science had gained

Figure 2.1 Life Magazine, 1962.

new attention as it could be framed as an issue of geopolitical concern and therefore motivated funding for defence purposes. This was most visible through the international scientific project *The International Geophysical Year 1957/58*, which had warranted funding for Charles Keeling's observations at Mauna Loa station in Hawaii, resulting in the now famous Keeling-curve. In this setting, climate science became a way to protect national interests and a reason for international cooperation on scientific progress. Weather control was seen as a "Great Prospect" that could aid states through climate modification, useful in areas such as warfare and agriculture.[22] As a Soviet textbook stated in 1960, the "end goal of meteorology" was to "artificially change the weather and the climate in a direction preferable for humans".[23] This framing would change during the following decade as the science was coupled with the growing issue of global environmental threats, and air pollution especially. Climate change was recognised as a problem for the survival of the human species, rather than a possibility to improve the nation 's security.[24] This shift, from an outspoken and well-funded science for the benefit of countries, to a science of risks which questioned the net benefits of industry and demanding regulation, ultimately exposed a rising conflict from the late 1960s onwards.

A new conflict emerges

Instead of erroneously focusing on climate change obstruction as emerging in 1988 or 1989, when the scientific community became more consistent regarding the phenomenon's reality and its anthropogenic origins, it is reasonable to highlight the combined threats to – and disruptions in – the Western capitalist system and industrial modernity that occurred during the late 1960s and early 1970s together with the simultaneous rise of global environmentalism.[25]

During this period, it became possible to detect and trace pollutants at a global scale through the emergence of computer models, calculations and data collection sites around the planet. The same technologies that enabled the gathering and analysis of huge amounts of data also aided the creation of scenarios and models to project possible futures.[26] As discussed above, climate science during the turn of the decade became associated with the environmental movement as CO_2 began to be acknowledged as a pollutant. As airborne pollution, which does not acknowledge the borders set up by nation states, was viewed as a key issue, environmental problems became or were made global.[27] These results triggered a concerted attempt to protect the industry and Western dominance from restrictions related to environmental protection and from demands from former colonies on reparations, independence and self-control. As Oreskes and Conway have so crucially detailed in their book *Merchants of Doubt*, this new conflict created rifts in the scientific community. Scientists who saw themselves as defenders of the market economy and the West increasingly sided with conservative politics and industry interests to oppose environmental regulations and eventually climate science.[28]

Back then, there was a growing concern that environmental destruction and resource depletion could lead to lost revenues and that this needed to be taken into consideration. The highly influential Club of Rome report *Limits to Growth*, published in 1972, was a response to such fears as it outlined a grim prospect of future resource scarcity in the tradition of Thomas Malthus,[29] and the need to protect and commodify these resources for the maintenance of the Western way of life. In the report, which emerged from the discussions in the Organisation for Economic Co-operation and Development (OECD),[30] one of the objectives was to ensure the possibility to continue the use of large amounts of energy (not least fossil fuels) while acknowledging global limits, or "to save capitalism from itself", as political theorist Romain Felli writes.[31] While not challenging the commodification of nature and the attached need of increasing the use of energy per se, it was at least an acknowledgement that "[t]he capitalist model is *based* on the environment".[32] Thus, oil companies needed, according to the report, to internalise environmental costs and protect reserves (including foreign ones) for the future. When grassroot environmental movements looked at the graphs, another insight arose: to put the planet before profits needed to include planned degrowth in fossil fuel energy and material used in industrial modern societies.[33] The fears among actors in the centre of global environmental politics were still mainly focused on overpopulation and the increased use of resources in the Global South. With the 1973 oil embargo, fears that the former colonies and areas under influence from USA and Britain would restrict international supply of fossil fuels seemed to materialise. The pushback to the ideas of limits to growth came both as full front attacks and as more subtle diversions.

The gospel of adaptation and proactive strategies from the fossil fuel industry

The response from the fossil fuel industry to the issue of climate change and environmental destruction has not just been one of outright rejection. Rather, responses have proactively aimed to protect the continued use of fossil fuels since the 1960s – and herein lie the origins to present-day obstruction.

Building on the growing popularity of future-oriented planning tools, the fossil fuel company Royal Dutch Shell (since 2022, Shell) began making use of such scenarios after 1967. In a revealing paper, historian Jenny Andersson shows how "the main role of scenarios was to invent and project new visions of the future territories of oil in a world beyond limits".[34] In short, the scenarios produced a liberty of action for the company as well as a rhetorical device against regulation at a time when the thresholds and limits of resource extraction were heatedly debated. This kind of incorporation of critique and myth-making has been identified as key strategies in corporate response to climate change during later periods as well.[35] Not only did companies such as Shell produce scenario tools that ensured their continued existence and portrayed them as guarantees of global welfare and stability, but they also helped finance research that would be used as evidence for the possibility of continued use of their products.

In the late 1960s, Shell's research branch, Shell Research Limited, helped finance the research conducted by James Lovelock, the father of the Gaia hypothesis.[36] This hypothesis posits the world as a living organism with the ability of self-regulation, a thesis that has also been highly inspirational for the environmental movement.[37] In the hands of the fossil industry, Gaia was strategically used to distract attention away from pollution stemming from fossil fuels and instead to see them as part of a natural adaptive system. The argument went that no matter the amount of emission, the world would stabilise and adapt.[38] During the 1970s and 1980s, this research was used to blur the line between emissions stemming from the use of fossil fuels and those stemming from "natural" factors such as the aforementioned volcanic eruptions. In other words, with the Gaia hypothesis, the fossil fuel industry had found a theory and claim about the workings of the atmosphere that could function as "an ethics of corporate scepticism".[39]

The use of scenario tools and the funding of specific science supported the perspective that fossil fuels were a vital part of the future modern world, or that fossil fuels were no more harmful than the natural carbon or methane cycles. This was not obstruction in its primary but secondary form. Proactive strategies were aimed to keep the option of continued use and extraction open, at a time when the evidence for human causes of climate change was still debated in the scientific community. What was created were different "technologies of legitimacy" for the fossil industry.[40] But even in the scientific community, few argued for a ban on fossil fuels. Instead, a culture of preparedness developed where the aim of states and scientists often was to continue the use of fossil fuels while, simultaneously, creating the scientific basis for future decisions.[41] It was a policy of delay.

Neoliberal countermovement and economics

As mentioned above, the environmental issue was one of three main changes that the Western economic system faced at the end of the 1960s and beginning of the 1970s. Responding to the demands from environmental movements, the recently decolonised states in the Global South, and the labour movement, a new countermovement was taking shape, building on the ideas of neoliberalism and neoconservatism.[42] As historian of science, Jeremy Walker states: "the spectre of a rising environmental movement was a catalyst for the mass-enrolment of transnational corporations in the neoliberal project".[43] Instead of the environment, industry and the corporate sphere started to position the economy and energy as the entities in need of protection.[44] The present-day movement against environmental regulation (including climate change policies) and global justice was to a large extent born at this time.[45]

The oil crisis of the early 1970s also meant that the price of oil sky-rocketed, which enabled some of the most profiled actors in the coming climate obstructionist movement to accumulate massive profits. Thus, companies such as Koch and Gulf Oil/Chevron could lay the infrastructural ground of the later decades' primary obstruction, through the creation of think tanks and foundations such as the Cato Institute.[46]

Parts of the industry and economic science responded to environmental regulation and the idea of limits with rejection, accusing environmentalists of promoting socialism.[47] The position of rejection lies within a deeply troublesome conflict between ecology and neoliberal thought that stems from the foundation of the two fields of knowledge. In neoliberal economic models, the very notion of land and finitude was excluded as the economic system was understood as an autonomous system with little or no relation to underlying material factors.[48] As ecology became a more and more respected and hailed science, and as the environmental movements grew during the 1960s, the conflict intensified. In the 1970s, neoliberal and neoconservative actors underpinned the rift between themselves and the environmental movements by insisting that the most radical branches were the true representatives of the environmental movement.[49] For example, in a 1973 piece for *The Public Interest* called "Capitalism, Socialism, Nihilism", neoconservative Irving Kristol wrote that environmentalists:

> are not really interested in clean air or clean water at all. What does interest them is modern industrial society and modern technological civilization, toward which they have profoundly hostile sentiments.[50]

Kristol and like-minded actors would help shape the polarised nature of climate politics in the decades to come through their later engagement in crucial obstructionist groups such as the American Enterprise Institute, Cato Institute, and the Heritage Foundation.

The concerted efforts to counter narratives of resource depletion and the unsustainable nature of the production system at the time consisted of attacks on what conservative actors purportedly called alarmist and doomist narratives.[51] This is a tactic we can recognise today as even IPCC reports are being accused of alarmism, explicitly invoking the debates of the 1970s.[52] Heartland Institute fellow Anthony Watts, who was instrumental in the spread of conspiracy theories in connection to the so-called Climategate back in 2009, stated in 2021: "It seems that climate disaster is always just five to 10 years away, but none of the predictions of climate doom have yet to come true".[53] Thus, Watts keeps reproducing a historical narrative that has almost become a trope within primary and secondary obstruction.[54]

These accusations of alarmism in the 1970s were not only targeting the radical left or environmentalism. The *Limits to Growth* report, produced at the very centre of economic power, was viciously opposed too. Instead of ideas of internalising the costs of environmental destruction in capitalism as the report had proposed, one strand within neoliberal thought developed an outright denial of any limitations to extraction and human interference with the physical world. The most profiled among the actors defending such a position was the U.S. economist and senior fellow at Cato Institute, Julian Simon. In Simon's vision, human ingenuity could overcome any obstacle and thus more humans would mean more "processing power" and a better chance to produce substitute resources, such as fossil fuels. Coupled with market forces and price signals, no firm action to address climate change was needed according to this logic.[55] Simon's intervention is interesting as it effectively challenged the Malthusian problem head on and implied that answers to future problems lie in the expansion and creative destruction of its causes.[56] Simon and others were crucial in developing a green growth counternarrative to environmentalists' positions. This was perhaps most forcefully made in the book *The Resourceful Earth: A Response to Global 2000* from 1984, co-written by Herman Kahn and sponsored by the Heritage Foundation, one of the most influential think tanks in the denial machine of the 1990s and 2000s.[57] While being most innovative in the USA, similar notions of technological fixes to adapt to a changing climate were also heard in the Soviet Union from officials such as E.K Fedorov, or in Sweden by public experts such as Tor Ragnar Gerholm.[58] The ideas developed by Simon have been called cornucupianism and point to the strategy among climate delayers in our present days to, for example, focus on currently used fuels as needed in a transition phase. This includes the idea of clean coal or the recent categorisation in the European Union taxonomy of fossil gas as green and the idea that we need to continue extraction to afford climate mitigation policies.[59] Thus, at the period when the need for action became apparent, a countermovement that actively discarded government action was gaining influence.[60]

Shifting responsibility, the politics of global warming

In addition to such outright deflection and hostility to the science and environmental movement's demands on industry, more subtle ways of redistributing responsibility were underway during the 1970s. Within the political sphere, climate change had, during the early 1970s, become one of many environmental risks associated with the modern society. In the preparations for the 1972 United Nations Conference on Human Environment in Stockholm, and in the plethora of future-oriented scenarios common at the time, the issue of climate change appeared.[61] To understand delay and inertia, two different cases from countries historically considered environmental forerunners, Sweden and Norway, are useful. The cases show how potentially important and simultaneously difficult the issue could become in a country that began its extraction of fossil fuels when climate science emerged and in one where the transition away from fossil fuels was already underway.

As a result of the oil crisis and a planned nuclear power expansion in Sweden, climate change became an argument against the growing anti-nuclear/pro-small-scale renewable movements in the early 1970s. The threat of continued oil-dependency was contrasted to the threat of reactor failures and the problems of waste storage by the Social Democratic Party. "It [CO_2 emissions] concerns half-life effects – and therefore about issues of responsibility for coming generations – fully comparable to radioactive waste".[62] But as the anti-nuclear movement gained foot, the argument of climate change became less useful for the political leaders and, thus, the issue was removed from national discussions and positioned not as a problem of fossil fuel production but of global CO_2 emissions.[63] Through such a reframing of the issue, the possibility to shift the blame away from territorial emissions to global emissions appeared and privileged policies focused on the demand rather than the supply side.[64] For example, when the Swedish meteorologist and later IPCC chairman Bert Bolin talked about climate change in the mid-1970s, he discussed a transition away from fossil fuels. In the 1980s, this idea had within the governmental investigations transformed into the need to fight total emissions.[65] This idea, that it is solely emissions on a global level that are important and that any one country cannot be held responsible for mitigation efforts, has continued to haunt climate policy and forcefully restricted any measures to regulate fossil fuel supply.[66] Such framings are in contemporary society actively countered in the calls to "keep it [fossil fuels] in the ground" from social movements such as *Ende Gelände* in Germany or at-risk-population protests such as the fight against the Dakota Access Pipeline.[67]

A similar process as the one in Sweden was seen in Norway during the 1970s. As Norway's extractive enterprise struck black gold at a time when the environmental threat of climate change became apparent, the need to shift responsibility became crucial. For Environmental Minister and later Prime Minister Gro Harlem Brundtland and the economist and Labour party's youth-wing leader Jens Stoltenberg, the issue of climate change presented an opportunity to wring the environmental issue from the hands of the Norwegian environmental movement and its radical branch, the deep ecologists.[68] These movements demanded domestic regulations and environmental protective legislation and, together with the indigenous Sámi population, organised massive protests against the expansion of hydropower. By promoting and developing the idea that domestic emissions could be compensated by investments in other parts of the world, what became known as Clean Development Mechanisms, Brundtland and Stoltenberg pioneered the idea that oil extraction and climate change policy could go hand in hand.[69] This was a strategy that effectively changed the narrative of environmental protection away from social or and ethical considerations into a question of mere technicality or an issue of economic calculation. When implemented in the Kyoto protocol 1997, this shift of responsibility did not go unnoticed and can be seen as a continuation of extractive practices by the Global North under the guise of carbon colonialism.[70]

The focus on the global arena for these two small Nordic countries served a dual purpose of acknowledging the character of the problem, simultaneously as it relieved individual states from obligations to mitigate or stop fossil fuel extraction – what in more recent times have been dubbed New denialism, when the analysis is done of Canada or the USA.[71] A similar logic could be seen in the temporal dislocation of actions evident in recent pledges for so-called net-zero goals or other temporal fixes.[72] Trivialising domestic efforts and framing climate change as an exclusively global issue further created the possibility to frame climate change policy as a collective action problem with potential free riders. The cost for any individual country to pursue mitigating efforts was presented as too high or risks of acting in advance were emphasised. Policy perfection was thus sought to counter such fears.[73]

Economy against decarbonisation

Strategies of delaying immediate action on the findings of both corporate and public climate science were enhanced by the promotion of neoclassical and neoliberal economic models in environmental policy later and ultimately institutionalised in laws, price mechanisms and markets.[74] Already in the 1960s, economists had begun to think about how to respond to the issue of uncertainty and complexity that issues like climate change posed. Drawing on neoclassical theory, economists started to calculate the costs and benefits of acting on environmental problems. For example, in 1968, two Swedish economists Assar Lindbeck and Erik Dahmén initiated a research project to formulate a "theory of investments under uncertainty" to better deal with environmental issues.[75] While this can be viewed, in some ways, as a response to the environmental concerns at the time, this newly found fascination for the future and uncertainty among oil majors and economists was also in accordance with the "gospel of adaptation".[76]

The formulation of climate change as a complex problem made market solutions built on the theories of Friedrich August Hayek, key thinker of neoliberalism, seem apt. As Hayek had argued, no single actor, state or individual, could hold sufficient information to make credible calculations on future needs and resources. The only mechanism able to perform such complex calculations was the market. Thus, governing environmental and climate protection through markets became the solution in the era of sustainable development.[77] This created two narratives that have limited decarbonisation efforts. First, it ensured the possibility that fossil fuel corporations could continue the extraction unless the market, through price signals, indicated them to shift. Second, the idea portrayed individual consumers and their use as the real culprits and actors driving or limiting greenhouse gas emissions. Indeed, the idea of consumer sovereignty was pioneered by Ludwig von Mises already in the 1920s but reused in relation to environmental issues in the 1960s and 1970s.[78] This strategy has been a constant companion to anti-regulatory climate policies since the early 1990s and used extensively in tandem with other obstruction strategies in more recent years.[79] In this circumstance, Philip Mirowski's portrayal of neoliberal thought in relation to climate change is of importance. Science denial, promotion of carbon trading and geoengineering are different strategies – with different temporal objectives – to reinforce the idea that markets will ultimately solve problems which arise from complex human/environment interactions.[80]

Since the late 1960s, economics had struggled with the issue of environmental problems, and by the 1970s and 1980s, the models from neoclassical and neoliberal economics had gained influence as frames to policy measures. Neoclassical models in turn have relied on cost-benefit calculations that consistently have downplayed the costs of climate change and calculated exaggerated costs for mitigation and decarbonisation. Most importantly, the economic models have relied on the idea that technological innovation will make climate policy cheaper in the future, thus limiting the benefit of acting now.[81] Furthermore, the goal of these policies and models, developed not the least by William Nordhaus since the 1970s, has never been to lower emissions as much as possible but to find the "optimal level".[82] As a result, the reliance on economic models and arguments has provided the rationale for not acting even though their intentions might have been different.[83] This economic style of reasoning has permeated environmental regulation since the 1960s and almost totally replaced previous notions of protecting the environment for the sake of ecological stability. Instead, only those factors that could be incorporated and accounted for in the economic system had relevance.[84] The economic assessments of environmental and climate policies as well as market logics have continuously been used to argue against environmental action and climate mitigation in the years after 1988 by maintaining the idea that climate action is always a cost.[85]

The PR campaign begins

As science became more and more certain and coherent about the issue of anthropogenic climate change, directed efforts to counter climate change action began to take shape in disparate ways. That is,

during the 1980s, the more general countermovement against environmental regulation that had emerged in the late 1960s began to make
use of experts and PR to question the acuteness of the issue as well as
targeting projections and reviews of the literature.[86] As we have seen
above, the use of PR in protecting industry actors from environmental
regulation had been used before for strategic reasons to, for example,
shift responsibility to consumption rather than production, but in the
late 1980s this narrative gained foot.[87] Some companies like car manu-
facturer Volvo had, since the 1970s, portrayed themselves as environmental forerunners to maintain their relevance and increase market
shares.[88] Already the early 1980s had seen the American Petroleum
Institute downplaying the risk of climate change in public campaigns
while simultaneously acknowledging the threat in internal communication.[89] Similarly, Exxon's internal research reports had acknowledged the reality of anthropogenic global warming early on. Their
research leader Henry Shaw said in 1981 that global warming could be
expected to reach a three-degree rise in a hundred years, but that there
was "time for an orderly transition to non-fossil fuel technologies if
restrictions on fossil fuels were deemed necessary".[90] However, instead
of acting on the science, companies like Exxon began promoting
advertorials that downplayed the risks.[91]

During the same period, PR companies and collaborative efforts
between industry, governments and environmental organisations
reframed the industry's position. Instead of just responding to attacks,
PR firms with fossil fuel majors and industry groups helped to change
the narrative to one where the industry groups weren't seen as culprits
but as leaders in the transition to a sustainable world. By taking com-
mand of the communication, the industry could dodge regulation by
taking on the responsibility for their environmental impact through
voluntary codes of conduct and self-regulation.[92] Therefore, fairs like
Ecology 1989 in Sweden (coordinated by Volvo) were celebrated as a
definitive turn and considered as evidence to corporations having realised the seriousness of climate change and other environmental problems.[93] Companies could, under the umbrella of green capitalism or
sustainable development, boast about their ambitious internal plans
but at the same time continue with only slight adjustments to their
businesses.[94] Through the organisations, such as the International
Chamber of Commerce companies like Shell could engage in "proactive leadership" on environmental issues, a leadership that was enacted
through the concept of sustainable development.[95]

But when this recent and fragile configuration of green business was
criticised, other measures were needed. If green communication was

one way to insulate against regulation, the other was more confrontative. Businesses targeted by environmental demands began setting up networks where information and tactics could be disseminated. One early network was the Atlas Network founded in 1981, which coordinated neoliberal organisations and think tanks which had emerged during the 1970s. It enabled strategies and ideas to travel between countries and thus provided a platform for the dissemination of tactics protecting industry against the demands for environmental protection.[96]

In 1988, there were plenty of evidence that many parts of society had come to see climate change, not just as a matter of fact but also a matter of concern, a problem that needed political action. This understanding was evident in the now-almost iconic testament before the U.S. senate by James Hansen after which he told a reporter: "It is time to stop waffling so much and say that the evidence is pretty strong that the greenhouse effect is here".[97] The seriousness is perhaps even more clearly seen in the inauguration of the IPCC and the beginning of a global collaboration of world governments in their work against global warming. It seemed that a transformation towards a fossil-free society was possible.

The implications of these shifts in public opinion and political awareness did not go unnoticed among the prime beneficiaries of fossil fuel extraction. Companies like Exxon responded, not by acting on the science they themselves had funded and produced, but by turning these efforts on its head when addressing the public.[98] A former environmental director of the French fossil fuel company Elf recollects how Exxon in 1984 warned other companies that a "collective response from the industry [was] required".[99] Accompanying the statements of severity and political action by world leaders, NGOs, social movements and recently elected Green parties, a more concerted obstructionist movement grew. The force of the anti-environmental movements and climate obstruction efforts became qualitatively different as climate change policy and science became the main threat for industry actors during the late 1980s.

Summary

In this chapter, we have provided an overview of the foundation of obstruction. By going over the political economy and history of fossil fuels and climate change, we have outlined a rough compilation of arguments that were fine-tuned and activated full force in the late 1980s up until the early 2020s, as some remain central today. Some of these arguments are worth going over again.

As this chapter has shown, efforts to obstruct decarbonisation are not only a result of directed attempts by the fossil fuel industry but of the world-making capacities of industries, economics and political doctrines combined. The result was not part of a diabolical and ominous plan but of a set of reactions to various threats faced by the industry and the political sphere. As such, several tactics developed which enabled shifting between strategies that insulated business-as-usual from disruptive interference. The fossil fuel industry cultivated the idea that oil was a necessity for a good life during the 20th century which has been a staple of climate obstructionists during more recent years. Together with the idea that, in the grand scheme of things, limiting extraction or creating effective national mitigation policies would be pointless, the fossil fuel industry has found(ed) a powerful narrative combining "whataboutism" and a sense that nothing could be different.[100] But as we have shown, this narrative of global responsibility and a focus on emissions, rather than production, was also pushed by national governments unwilling to act. The blame for CO_2 emissions has not only been directed at the global scale. Instead, climate change as a complex issue provided an argument for the introduction of market measures and individual responsibility regarding climate change. Underlying this is a continuing emphasis on adaptation to climate change rather than mitigation, and as such these strategies are best understood as delay or secondary obstruction. In this chapter, we have also depicted the rise of the neoliberal and neoconservative movements forcefully rejecting environmental movements and ecological science. Through notions of cornucopia and technological optimism, this countermovement provided a critique of the idea of limits to growth and the rationales for not transitioning away from fossil fuels. With the profits from the high oil prices in the 1970s, they created and expanded the organisational structures that the climate change countermovement could thrive on and use as a template for future operations. And these developments are what we turn to next.

Notes

1 Weart, S. (2016) "Global warming: how skepticism became denial", *Bulletin of the Atomic Scientists*, 67(1), pp.41–50. https://doi.org/10.1177/0096340210392966.
2 Mann, M.E. (2021) *The New Climate War: The Fight to Take Back Our Planet*. London: Scribe; Aronczyk, M. and Espinoza, M.I. (2022) *A Strategic Nature: Public Relations and the Politics of American Environmentalism*. Oxford: Oxford University Press. https://doi.org/10.1093/oso/9780190055349.001.0001.

3 McGoey, L. (2019) *The Unknowers: How Strategic Ignorance Rules the World*. London: Bloomsbury Publishing; Proctor, R.N. and Schiebinger, L. (2008) *Agnotology: The Making and Unmaking of Ignorance*. Stanford: Stanford University Press; Verburgt, L.M. (2020) "The history of knowledge and the future history of ignorance", *KNOW: A Journal on the Formation of Knowledge*, 4(1), pp.1–24. https://doi.org/10.1086/708341; Hultman, M., Kall, A.S. and Anshelm, J. (2021) *Att ställa frågan. Att våga omställning. Birgitta Hambraeus och Birgitta Dahl i den svenska energi- och miljöpolitiken 1971-1991*. Lund: Arkiv förlag.

4 Malm, A. (2016) *Fossil Capital: The Rise of Steam Power and the Roots of Global Warming*. London: Verso.

5 Mitchell, T. (2011) *Carbon Democracy. Political Power in the Age of Oil*. London: Verso.

6 Ibid.

7 Ibid.

8 Steffen, W. *et.al.* (2015) "The trajectory of the Anthropocene: The Great Acceleration", *The Anthropocene Review*, 2(1), pp.81–98. https://doi.org/10.1177/2053019614564785; McNeill, J.R. (2016) *The Great Acceleration*. Cambridge, MA: Harvard University Press.

9 Huber, M.T. (2013) *Lifeblood: Oil, Freedom, and the Forces of Capital*. Minneapolis: University of Minnesota Press.

10 FreeEntertainment11 (2013) "Two Fords Classic Car Commercials (1950's)", *YouTube*, 1 April. Available at: https://www.youtube.com/watch?v=kmzlbkhyCJ4 [Last accessed March 29, 2022].

11 "Intresset för kol som alternativ till oljan större i våras än idag", *Göteborgs-Posten*, November 4, 1977.

12 Malm, A. and the Zetkin collective (2021) *White Skin, Black Fuel: On the Danger of Fossil Fascism*. London: Verso. See also in Chapter 4 on the use of cars among far-right discourses of obstruction.

13 Bell, A. (2021) *Our Biggest Experiment: A History of the Climate Crisis*. London: Bloomsbury Sigma.

14 Weart, S. (2016) "Global warming: how skepticism became denial"; Weart, S. (2008 [2003]) *The Discovery of Global Warming*. Cambridge, MA: Harvard University Press.

15 "Naturvetenskapliga kåserier", *Arbetet*, March 21, 1904.

16 Pasek, A (2021) "Carbon vitalism: life and the body in climate denial", *Environmental Humanities*,13(1),pp.1–20.https://doi.org/10.1215/22011919-8867175.

17 Buch, K. (1951) "Kolsyra och klimat", *Teknisk tidskrift*, 81.

18 Doose, K. and Oldfield, J. (2018) "Natural and anthropogenic climate change understanding in the Soviet Union, 1960s–1980s" in Poberezhskaya, M. and Ashe, T. (eds) *Climate Change Discourse in Russia*. Abingdon, Oxon: Routledge, pp.26–27.

19 Franta, B. (2018) "Early oil industry knowledge of CO2 and global warming", *Nature Climate Change*, 8, pp.1024–1025. https://doi.org/10.1038/s41558-018-0349-9.

20 Ibid.

21 Ibid., p.1024.

22 Schubert, J. (2021) *Engineering the Climate – Science, Politics, and Visions of Control*. Manchester: Mattering Press, p.223. https://doi.org/10.28938/9781912729265.

23 Skliarorv, V.M. (1960) Meteorologiia i Metoeorologicheskie Nabliudeniia (Obshchedostupnyi kurs) Gidrometeoizdat, Lenigrad, quoted in Doose, K. and Oldfield, J. (2018) "Natural and anthropogenic climate change understanding in the Soviet Union, 1960s–1980s".

24 Howe, J.P. (2014) *Behind the Curve: Science and the Politics of Global Warming.* Seattle: University of Washington Press, p.43.

25 Aronczyk, M. and Espinoza, M.I. (2022) *A Strategic Nature.*

26 Edwards, P.N. (2010) *A Vast Machine: Computer Models, Climate Data, and the Politics of Global Warming.* Cambridge, MA: University of Washington Press.

27 Buns, M.A. (2020) "Green internationalists, nordic cooperation, 1967-1988" (Doctoral thesis), Univeristy of Oslo; Felli, R. (2021) *The Great Adaptation – Climate, Capitalism and Catasrophe.* London: Verso, p.25; Heidenblad, D.L. (2021) *The Environmental Turn in Postwar Sweden: A New History of Knowledge.* Lund: Lund University Press.

28 Oreskes, N. and Conway, E.M. (2010) *Merchants of Doubt: How a Handful of Scientists Obscured the Truth on Issues from Tobacco Smoke to Global Warming.* London: Bloomsbury.

29 Therefore, this trend is usually described as Neo-Malthusian. Malthus argued that without regulations and control populations would outgrow the supply of food. Malthusianism is therefore often used as a slur or pejorative term against arguments of overpopulation or resource scarcity.

30 Schmelzer, M. (2017) "'Born in the corridors of the OECD': the forgotten origins of the Club of Rome, transnational networks, and the 1970s in global history", *Journal of Global History*, 12(1), pp.26–48. https://doi.org/10.1017/S1740022816000322.

31 Felli, R. (2021) *The Great Adaptation*, p.25.

32 Wright, C. and Nyberg, D. (2015) *Climate Change, Capitalism, and Corporations.* Cambridge: Cambridge University Press, p.34. https://doi.org/10.1017/CBO9781139939676.

33 Ekberg, K. and Hultman, M. (2021) "A question of utter importance: the early history of climate change and energy policy in Sweden, 1974–1983", *Environment and History*, 8. https://doi.org/10.3197/0967340 21X16245313030028.

34 Andersson, J. (2020) "Ghost in a Shell: the scenario tool and the world making of Royal Dutch Shell", *Business History Review*, 94(4), pp.729–751. https://doi.org/10.1017/S0007680520000483.

35 Wright, C. and Nyberg, D. (2015) *Climate Change, Capitalism, and Corporations.*

36 Aronowsky, L. (2021) "Gas Guzzling Gaia, or: a prehistory of climate change denialism", *Critical Inquiry*, 47(2), pp.306–327. https://doi.org/10.1086/712129.

37 Lovelock, J. (2016 [1972]) *Gaia: A New Look at Life on Earth.* Oxford: Oxford University Press.

38 Aronowsky, L. (2021) "Gas Guzzling Gaia, or: a prehistory of climate change denialism", *Critical Inquiry*, 47(2), pp.306–327. https://doi.org/10.1086/712129.

39 Ibid., p.315.

40 Technology of legitimacy is a concept developed by Melissa Aronczyk and Maria I. Espinoza and refers both to the ways in which actors secure

legitimacy for one viewpoint and how they structure what is conceivable. Aronczyk, M. and Espinoza, M.I. (2022) *A Strategic Nature*, pp.9–10.

41 Felli, R. (2021) *The Great Adaptation*, chapter 1, pp.17–53.

42 Boynton, Alex. (2015) "Confronting the environmental crisis? Anti-environmentalism and the transformation of conservative thought in the 1970s" (Doctoral thesis), University of Kansas. See also Layzer, J.A. (2012) *Open for Business: Conservatives' Opposition to Environmental Regulation*. E-book. Cambridge, MA: MIT Press.

43 Walker, J. (2020) *More Heat than Life: The Tangled Roots of Ecology, Energy, and Economics*. Singapore: Palgrave Macmillan, p.13.

44 Aronczyk, M. and Espinoza, M.I. (2022) *A Strategic Nature*, chapter 3, pp.71–95.

45 Though Walker importantly notices that the connections between fossil fuel corporations and the neoliberal thought collectives probably extends further back in time. Walker, J. (2020) *More Heat than Life*, p.29.

46 Mitchell, T. (2011) *Carbon Democracy*, pp.197–198; Oreskes, N. (2018) "The scientific consensus on climate change: how do we know we're not wrong?" in Lloyd, E.A. and Winsberg, E. (eds) *Climate Modelling: Philosophical and Conceptual Issues*. Cham: Springer International Publishing, p.52.

47 Felli, R. (2021) *The Great Adaptation*; Oreskes, N. and Conway, E.M. (2010) *Merchants of Doubt*, pp.246–262.

48 Walker, J. (2020) *More Heat than Life*, pp.157–158. As Timothy Mitchell argues the establishment of economy as a field of knowledge and science meant the privilege of models that was less about nature and resources and more about prices and money.

49 Boynton, A. (2015) "Formulating an anti-environmental opposition: neoconservative intellectuals during the environmental decade", *The Sixties*, 8(1), pp.1–26.

50 Quoted in Boynton, A. (2015) "Formulating an anti-environmental opposition", p.12. As Boynton writes these neoconservative actors would go on working in the very think tank networks that were key in the manufacturing of primary obstruction such as CATO and Heritage Foundation.

51 Gerholm, T.R. (1972) *Futurum exaktum: den tekniska utmaningen*. Stockholm: Aldus/Bonnier; Kahn, H. and Wiener, A.J. (1967) *The Year 2000: A Framework for Speculation on the Next 33 Years*. New York: Macmillan.

52 This while the IPCC reports have historically been cautious in their statements regarding the pace and effects of climate change, see: Brysse, K., Oreskes, N., O'Reilly, J. and Oppenheimer, M. (2013) "Climate change prediction: erring on the side of least drama?", *Global Environmental Change*, 23(1), pp.327–337. https://doi.org/10.1016/j.gloenvcha.2012.10.008

53 Taylor, J., Burnett, H.S. and Watts, A. (2021) "Press release: heartland Institute reacts to 'alarmist' UN IPCC climate report", Heartland Institute, 9 August. Available at: https://www.heartland.org/news-opinion/news/press-release-heartland-institute-reacts-to-alarmist-un-ipcc-climate-report [Last accessed March 29, 2022].

54 Ekberg, K. and Pressfeldt, V. (forthcoming 2022) "A road to denial: climate change and neoliberal thought in Sweden, 1988–2000", *Contemporary European History*.

55   Vettese, T. (2019) "Limits and cornucupianism: a history of neo-liberal environmental thought 1920-1970"(Doctoral thesis), New York University.
56   Vettese, T. (2019) "Limits and cornucupianism"; Wright, C. and Nyberg, D. (2015) *Climate Change, Capitalism, and Corporations*.
57   Simon, J.L. and Kahn, H. (1984) *Resourceful Earth: A Response to Global 2000*. Oxford: Blackwell Pub.; Simons and Kahn's thinking was exported around the world, one evidence of this is the Swedish Employers Association's (SAF) sponsoring of Kahn's and Gerholm's similar report: Gerholm, T.R. and Kahn, H. (1984) *En bättre framtid*. Stockholm: Sv. arbetsgivarefören.
58   Doose, K. and Oldfield, J. (2018) "Natural and anthropogenic climate change understanding in the Soviet Union, 1960s–1980s", pp.26–27; Ekberg, K. and Pressfeldt, V. (forthcoming 2022) "A road to denial".
59   European Comission (2022) "Press release: EU taxonomy: Commission presents complementary climate delegated act to accelerate decarbonisation", 2 February. Available at: https://ec.europa.eu/commission/presscorner/detail/en/ip_22_711 [Last accessed March 29, 2022].
60   Naomi Klein has famously called the rise of political awareness of climate change in the same period of neoliberalism's breaktrough a "epic case of bad historical timing" in Klein, N. (2015) *This Changes Everything: Capitalism vs. the Climate*. New York: Simon and Schuster, p.201. See also: Layzer, J.A. (2012) *Open for Business*; Parr, A. (2014) *The Wrath of Capital: Neoliberalism and Climate Change Politics*. Columbia University Press; Schneider, J., Schwarze, S., Bsumek, P.K. and Peeples, J. (2016) *Under Pressure: Coal Industry Rhetoric and Neoliberalism*. Springer. https://doi.org/10.1057/978-1-137-53315-9.
61   SMIC (1971) *Inadvertent Climate Modification – Report of the Study of Man's Impact on Climate (SMIC)*. Cambridge, MA: MIT Press.
62   "Fälldin och energin" Government's office, Press secretary, 7 February 1975, p.1 (a previous version exists issued by Tage Levin 10 December 1974 to TGP) Box A14:008, Appendix, Energifrågan. Olof Palme, Swedish Labour Movement's Archives and Library.
63   Ekberg, K. and Hultman, M. (2021) "A question of utter importance".
64   Green, F. and Denniss, R. (2018) "Cutting with both arms of the scissors: the economic and political case for restrictive supply-side climate policies", *Climatic Change* 150(1), pp.73–87. https://doi.org/10.1007/s10584-018-2162-x.
65   Ekberg, K. and Hultman, M. (fast track 2021) "A question of utter importance, the early history of climate change and energy policy in Sweden 1974-1983," *Environment and History*. https://doi.org/10.3197/0967340 21X16245313030028.
66   Buck, H.J. (2021) *Ending Fossil Fuels: Why Net Zero Is Not Enough*. Verso; Aykut, S.C. and Castro, M. (2017) "The end of fossil fuels?: Understanding the partial climatisation of global energy debates" in Aykut, S.C., Foyer, J. and Morena, E. (eds) *Globalising the Climate*. E-book: Routledge.
67   Estes, N. (2019) *Our History Is the Future: Standing Rock versus the Dakota Access Pipeline, and the Long Tradition of Indigenous Resistance*. London: Verso; McAdam, D. *et al.* (2010) "'Site fights': explaining opposition to pipeline projects in the developing world," *Sociological Forum*, 25(3).

https://doi.org/10.1111/j.1573-7861.2010.01189.x; https://onlinelibrary. wiley.com/doi/abs/10.1111/j.1573-7861.2010.01189.x; Sander, H. (2017) "Ende Gelände: Anti-Kohle-Proteste in Deutschland," *Forschungsjournal Soziale Bewegungen*, 30(1). https://doi.org/10.1515/fjsb-2017-0004; https:// doi.org/10.1515/fjsb-2017-0004.

68 The deep ecologists were a radical environmental group whose values were spelled out and defined by Arne Naess in the book *Ekologi, samhälle och livsstil* published in 1973.

69 Anker, P.J. (2018) "A pioneer country? A history of Norwegian climate politics", *Climatic Change*, 151(1), pp.29–41. https://doi.org/10.1007/s10584-016-1653-x.

70 Bachram, H. (2004) "Climate fraud and carbon colonialism: the new trade in greenhouse gases", *Capitalism Nature Socialism*, 15(4), pp.5–20. https:// doi.org/10.1080/1045575042000287299.

71 Daub, S. *et al.* (2020) "Episodes in the new climate denialism" in Carroll, W.K. (ed) *Regime of Obstruction: How Corporate Power Blocks Energy Democracy*. AU Press. https://doi.org/10.15215/aupress/9781771992893.01; Felli, R. (2021) *The Great Adaptation*, pp.59–60.

72 Carton, W. (2019) "'Fixing' climate change by mortgaging the future: negative emissions, spatiotemporal fixes, and the political economy of delay", *Antipode*, 51(3), pp.750–769. https://doi.org/10.1111/anti.12532; Buck, H.J. (2021) *Ending Fossil Fuels*.

73 Lamb, W.F. *et al.* (2020) "Discourses of climate delay", *Global Sustainability*, 3, e17; Doose, K. and Oldfield, J. (2018) "Natural and anthropogenic climate change understanding in the Soviet Union, 1960s–1980s", pp.35–49. https://doi.org/10.1017/sus.2020.13.

74 As such neoliberal and neoclassical economy has tended to view the "economy" as a separate sphere only dependent on the internal logic of the system. See: Mitchell, T. (2011) *Carbon Democracy*; Walker, J. (2020) *More Heat than Life*. For a critique of neoclassical economics ability to mitigate climate change see: Keen, S. (2020) "The appallingly bad neoclassical economics of climate change", *Globalizations*, 18(7), pp.1149–1177. https://doi.org/10.1080/14747731.2020.1807856.

75 Heidenblad, D.L. (2021) *The Environmental Turn in Postwar Sweden*, p.104.

76 Felli, R. (2021) *The Great Adaptation*, pp.8–9.

77 Ekberg, K. and Pressfeldt, V. (forthcoming 2022) "A road to denial"; Wright, C. and Nyberg, D. (2015) *Climate Change, Capitalism, and Corporations*.

78 Olsen, N. (2020) "Ludwig von Mises, the idea of consumer democracy and the invention of neoliberalism", *The Tocqueville Review*, 41(2), pp.43–64. https://doi.org/10.3138/ttr.41.2.43; Mann, M.E. (2021) *The New Climate War*.

79 Supran, G. and Oreskes, N. (2021) "Rhetoric and frame analysis of ExxonMobil's climate change communications", *One Earth*, 4(5), pp.696–719. https://doi.org/10.1016/j.oneear.2021.04.014; Ekberg, K. and Pressfeldt, V. (forthcoming 2022) "A road to denial"; Mckie, R.E. *et al.* (2021) "The Mises Institute Network and Climate Policy. 9 findings", *CSSN Research Report* 2; Mann, M.E. (2021) *The New Climate War*.

80 Mirowski, P. (2013) *Never Let a Serious Crisis Go to Waste : How Neoliberalism Survived the Financial Meltdown.* London: Verso, pp.334–342.
81 Carton, W. (2019) "'Fixing' climate change by mortgaging the future". By using what economist talk about as a discount rate.
82 Randalls, S. (2011) "Optimal climate change: economics and climate science policy histories (from Heuristic to Normative)", *Osiris,* 26(1), pp. 224–242; Keen, S. (2020) "The appallingly bad neoclassical economics of climate change"; Masini, F. (2021). "William Nordhaus: a disputable nobel [prize]? Externalities, climate change, and governmental action", *The European Journal of the History of Economic Thought,* 28(6). https:// doi.org/10.1080/09672567.2021.1963798.
83 Stoddard, I. *et al.* (2021) "Three decades of climate mitigation: why haven't we bent the global emissions curve?", *Annual Review of Environment and Resources,* 46(1), pp.653–689. https://doi.org/10.1146/annurev-environ-012220-011104.
84 Berman, E.P. (2022) *Thinking Like an Economist: How Efficiency Replaced Equality in US Public Policy.* E-book: Princeton University Press, pp.6–10.
85 Franta, B. (2021) "Weaponizing economics: big oil, economic consultants, and climate policy delay", *Environmental Politics.* https://doi.org/10.1080/ 09644016.2021.1947636; Ekberg, K. and Pressfeldt, V. (forthcoming 2022) "A road to denial".
86 Oreskes, N. and Conway, E.M. (2010) *Merchants of Doubt,* pp.175–183.
87 Mann, M.E. (2021) *The New Climate War;* Aronczyk, M. and Espinoza, M.I. (2022) *A Strategic Nature.*
88 Bergquist, A.-K. and Näsman, M. (2021) "Safe before green! The greening of Volvo cars in the 1970s–1990s", *Enterprise & Society,* pp.1–31. https:// doi.org/10.1017/eso.2021.23.
89 Franta, B. (2018) "Early oil industry knowledge of CO2 and global warming".
90 Banerjee, N. (2015) "More Exxon documents show how much it knew about climate 35 years ago", *Inside Climate News,* 1 December.
91 Supran, G. and Oreskes, N.J. (2017) "Assessing ExxonMobil's climate change communications (1977–2014)", *Environmental Research Letters,* 12, 084019. https://doi.org/10.1088/1748-9326/aa815f; Boon, M. (2019) "A climate of change? The oil industry and decarbonization in historical perspective", *Business History Review,* 93(1), pp.101–125. https://doi. org/10.1017/S0007680519000321.
92 Aronczyk, M. and Espinoza, M.I. (2022) *A Strategic Nature;* Wright, C. and Nyberg, D. (2015) *Climate Change, Capitalism, and Corporations.*
93 The natural step was also the name of a consultancy focused on greening enterprise. Ekberg, K. and Pressfeldt, V. (forthcoming 2022) "A road to denial".
94 Bergquist, A.-K. and David, T. (2022) "Beyond limits to growth!: collaboration between the international business and United Nations in shaping global environmental governance", *Les Cahiers de l'IEP, IEP Working Paper Series,* 80, Université de Lausanne; Jones, G. (2017) *Profits and Sustainability: A History of Green Entrepreneurship.* Oxford: Oxford University Press. https://doi.org/10.1093/oso/9780198706977.001.0001.

95 Bergquist, A.-K. and David, T. (2022) "Beyond limits to growth!: collaboration between the international business and United Nations in shaping global environmental governance", *Les Cahiers de l'IEP, IEP Working Paper Series*, 80, Université de Lausanne; Jones, G. (2017) *Profits and Sustainability: A History of Green Entrepreneurship*. Oxford University Press, pp.27–28. https://doi.org/10.1093/oso/9780198706977.001.0001.

96 DeSmog. "Atlas Network (Atlas Economic Research Foundation)". Available at: https://www.desmog.com/atlas-economic-research-foundation/ [Last accessed April 21, 2022].

97 Shabecoff, P. (1988) "Global warming has begun, expert tells senate", *New York Times*, 24 June.

98 Supran, G. and Oreskes, N.J. (2017) "Assessing ExxonMobil's climate change communications (1977–2014)".

99 Quoted from Bonneuil, C., Choquet, P.L. and Franta, B. (2021) "Early warnings and emerging accountability: total's responses to global warming, 1971–2021", *Global Environmental Change*, 71, 102386. https://doi.org/10.1016/j.gloenvcha.2021.102386.

100 In the tradition of Fisher such a position could be called "fossil fuel realism", see: Fisher, M. (2009) *Capitalist Realism: Is There no Alternative?* Lanham: John Hunt Publishing; Kornbluh, A. (2020) "Climate realism, capitalist, and otherwise", *Mediations*, 33(1–2), pp.99–118.

3   Organised climate obstruction

Introduction

As discussed in the previous chapter, incumbents and vested interests
of the fossil fuel industry have had incentives to block environmental
knowledge and policies considered as threats to their goods and businesses already during the mid-20th century. Back then, their obstruction was delaying action to decrease emissions of greenhouse gases
warming up the planet.[1] However, towards the end of the 1980s, and in
light of increasing scientific consensus, successes by Green political
parties and public and institutional awareness and willingness to act
against global climate change, fossil fuel companies were at a cross-
roads. They had what could be considered as the best available science
on climate change at the time. But instead of acting upon this science
and envisioning a society beyond industrial modernity powered by fossil fuels, these companies have been at the centre of doubling total
emissions of greenhouse gases since then. Indeed, globally, four out of
five units of energy sold and used in 2021 still came from gas, coal and
oil.[2] What happened?

Understanding the primary obstruction undertaken by the fossil
fuel industry itself is central to comprehend climate change denialism
at large. Companies such as ExxonMobil actively decided on a strategy of questioning their own scientific findings through manufacturing a narrative of literal denial.[3] An influential fossil-fuelled denial
machine assembled from 1988 onwards, mainly funded by extractive
companies and consisting of (neo)conservative and neoliberal think
tanks, coalitions of large scale industries, right-wing legacy and digital
media, and later, networked influencers, creating a propaganda system. The latter has spread its messages mostly to conservative politicians and company leaders with specific demographics (e.g. elderly
men) being particularly open to such ideas.[4]

DOI: 10.4324/9781003181132-3

Simultaneously, the shift to green corporate PR had begun and the fossil fuel industry and companies dependent hereof worked hard to portray themselves not as obstructers to an environmentally sounder society but as a prerequisite for it. Thus, companies such as British Petroleum and governments in Europe such as Norway took a slightly different obstructionist path.[5] A large body of research,[6] primarily looking at European and global climate politics, has studied what we call secondary obstruction in the form of the marketisation of carbon emissions, consumer responsibility, dismissing per capita accountability of local and national actions. Such obstruction has been carried out via blame games directed to whole nations, setting up carbon markets and supporting unrealistic hopes connected to large-scale terraforming technologies, such as Carbon Capture and Storage.[7] Such a way of dealing with the climate emergency would have, maybe, been an effective strategy if it had been forcefully and quickly implemented in the 1990s instead of greenwashing the continued increase in the extraction of fossil fuels as well as the doubling of the cumulative emissions of greenhouse gases in 2020 compared to pre-1990.[8] In 2022, we are living with the failure of such ecomodernism tropes and with the successful implementations of strategies of secondary obstruction.[9]

Assembling the climate denial machine

The years 1988–1990 marked the start of the establishment of a primary obstruction apparatus, namely the climate denial machine. Manufacturing doubt, in full contradiction with the fossil industry's own knowledge on climate science produced by their "in-house" experts,[10] was only one of several strategies used by actors with vested interests in fossil fuel production from late 1980s onwards. Obstructionist organisations used all means possible to spread distrust of climate science and policy, from publishing books, articles and, interviews, to drafting policy documents and organising government hearings.[11]

In the USA, three influential actors made important moves during these years. First, the conservative think tank George C. Marshall Institute (which would later become the CO_2 Coalition), founded in the mid-1980s, started to publish articles and reports on climate change. Many of them, such as *Global Warming: What Does the Science Tell Us?* (1990), were co-written by its three founding members, Frederick Seitz, William Nierenberg and Robert Jastrow, and literally denied the existence of anthropogenic climate change, arguing that a warmer world was caused by natural fluctuations due to solar activities.[12]

That same year, ExxonMobil, one of world's five largest companies overall at the time by revenue, employees and market value, also produced strategies to obstruct climate policies by similarly emphasising the uncertainty of the scientific conclusions on the reality of climate change. Figure 3.1 containing excerpts of Exxon's internal memos from the late 1980s until the late 1990s, exemplifies this combination of strategies: casting doubt and confusion by questioning science and pretending to care about the environment.[13]

The third key move was accomplished when the fossil industry effectively joined forces with other extractive sectors in the lobby organisation Global Climate Coalition which brought together primarily companies from the utility, steel/rail and energy sectors from around the globe to obstruct climate action.[15] An organised campaign was set up, opposing any international action on climate change. This undermined

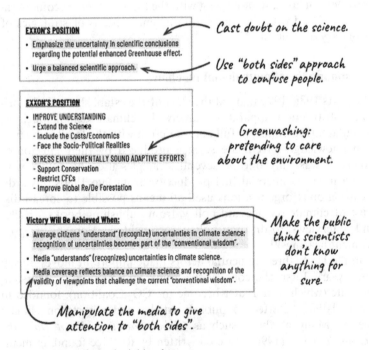

Figure 3.1 Notes from denial land.[14]

Source: reproduced from Cook *et al.* 2019.

the findings of their own internal research as well as climate scientists in academia.[16]

The U.S. denial machine as a whole, its components and their inter-relations are nowadays well understood. This concept, one developed by Riley Dunlap and Aaron McCright, describes the intricate connections between actions of obstruction that appeared independent but which were in fact well-coordinated. To describe this web, Robert Brulle speaks of a "climate change countermovement", thus emphasising the shared collective identity, aims of the actors involved and their coordinated struggle against political opponents.

The funding and creation of primary obstruction

It is difficult to obtain precise information on how much the sources of primary obstruction such as the fossil fuel industry, corporate America and conservative foundations spend on preventing climate policy and spreading doubt. Many of the organisations try to keep their contributors' secret. However, as it is possible to trace funding in the USA, thanks to the legislation on transparency of donation, it has been shown that a large part of the money comes from foundations that are linked to the fossil fuel industry. Mapping this money spent on lobbyism, Brulle went through the public registration forms that tax-exempt organisations must submit and compared them with information from a foundation database for the years 2003–2010. In this way, he was able to reconstruct the flow of money from conservative foundations to the climate denial machinery. The four largest funders were the Koch brothers, Exxon, the Scaife foundations and the conservative Lynde and Harry Bradley foundation.[18] In a follow-up study, Brulle discussed how, during the years 2000–2016, the fossil fuel industry in the USA spent ten times more money on lobbying to influence climate legislation than the environmental movement and the renewable energy industry combined. During the previous decade, 1990–1999, the fossil fuel industry was judged to have an even significantly larger advantage because the renewable industry was then economically weaker.[19] However, it is only part of the funding of the climate denial machine that Brulle managed to identify in these two studies. The rest, so-called dark money, poses a democratic problem as the public does not know what are the politico-economic interests that finance a particular organisation.[20] In fact, since 2006 the opaqueness has become even more thorough in the USA – more similar to the secrecy of money flows connected to Conservative Think Tanks (CTTs) in Europe.[21]

Brulle's analysis shows that contributions from two funds known as the Donors Trust and the Donors Capital Fund began to rise sharply that year, while the money from the other large funds decreased. Officially, for example, Exxon has completely stopped financing climate change denial, but it is difficult to know if this is really the case. Donors Trust and Donors Capital Fund are funds that do not manage any money themselves. They are only intermediaries of grants that allow the real donors/financers to remain anonymous. As the Donors Trust states, their mission is "to ensure the intent of donors dedicated to the ideals of limited government, personal responsibility and free enterprise".[22] Grants provided through these two funds in 2010 accounted for almost 25 per cent of the total to organisations part of the USA denial machine. Throughout the period 2003–2010, $78 million were channelled through the Donors Trust and the Donors Capital Fund to organisations that spread climate change denial.[23] Through the use of think tanks, astroturf organisations and front groups, what fuels the denial machinery can remain hidden – but as shown in the previous chapter, several of these organisations were funded by the revenues from oil in the 1970s. That is, through an anonymising middleman, traces of the fossil fuel industry and the conservative foundations can be more easily swept away.

Conservative Think Tanks

The receiver and user of the money that comes upstream from the sources (fossil fuel industry etc.) are most often conservative think tanks. As we saw in Chapter 2, the revenues from the high price of oil in the 1970s helped fund such enterprises and in the 1980s and 1990s further attempts to create collaborative structures among which think tanks took shape. A broad range of think tanks, in some cases founded by the extractive industry, began to spread doubt about climate science and to generally obstruct climate change mitigation. Collaborations, for example the previously mentioned Atlas Network, could coordinate and disseminate responses and policy suggestions.[24] The similarity in messages from think tanks and other actors in the denial machine speaks to the diffusion of tactics and talking points in similar networks.[25] Conservative and neoliberal think tanks in the USA, Australia, the UK and Canada, with close ties to the fossil fuel industry have been crucial component in obstructing and opposing climate policies.[26] They are intermediaries of money and ideology. In this section, we focus on five of the most influential actors known to this date: the George C. Marshall Institute, the Heartland Institute, the Cato

Institute, the Scaife Foundations and the Competitive Enterprise Institute (CEI).

One think tank which gained a leading role in the opposition to climate science was the neoliberal George C. Marshall Institute, founded in 1984. It would later become the CO_2 Coalition, whose leader, William Happer, was later one of Donald Trump's most important advisers. The think tank was founded by Frederick Seitz, Robert Jastrow and William Nierenberg, three physicists described by the historians of science Naomi Oreskes and Eric M. Conway as political hawks and sons of the Cold War in *Merchants of Doubt: How a Handful of Scientists Obscured the Truth on Issues from Tobacco Smoke to Global Warming*. It was named after the general who was the architect behind the Marshall Plan with the aim of spreading neoliberal ideas over the world. As written earlier, Seitz, Jastrow and Nierenberg co-wrote articles and reports, one of them being the institute's report on climate change published in 1990. William Nierenberg had a few years earlier acknowledged climate change when part of the American Academy of Sciences' work on the climate report commissioned by the Reagan administration. However, in the Marshall report, he argued that global warming was not related to larger greenhouse gas emissions. Instead, it was the sun that was behind the planet's rise in average temperature. The report was well received by the Republican Bush administration (1989–1993), which got a reason to justify its climate inaction in line with earlier standpoints on environmental questions, in contrast to what it had promised during the election campaign.

Another organisation, the Heartland Institute is perhaps the most explicit of the think tanks fighting climate science and policy in the USA.[27] Still much active, its activities are aimed at influencing decision-makers and public opinion. The Heartland Institute funds pseudoscience, produces reports, writes articles or is in direct contact with journalists, politicians and authorities. It was in the early 21st century that Heartland expanded its role in the climate denial machine with *Nongovernmental International Panel on Climate Change* (NIPCC), whose main author was Fred Singer, a physicist affiliated with the Heartland Institute and a central figure of climate change denialism.[28] It is not only the name that is confusingly similar to the UN's climate panel IPCC, but also the layout of their reports. These reports pretend to have been written by lead authors and co-authors and being peer-reviewed just like UN IPCC reports. However, as the warnings in the IPCC reports become increasingly strong, the NIPCC continued to engage in primary obstruction by undermining evidence regarding human impacts on the climate and claiming that the burning of fossil fuels does not lead to any negative consequences. A recent example is how NIPCC on October 5, 2018 (the same day that the IPCC report was released) stated that humans, animals and plants all have a better life if greenhouse gases levels in the atmosphere continue to rise.

Furthermore, David and Charles Koch, whose company Koch Industries was one of the largest oil companies in the USA, became influential in obstructionist circles. They started the Cato Institute, a think tank that has fought against all forms of government market regulation in all sectors, from education to healthcare and, not least, regarding the climate issue where they have been active in spreading denial. The big money coming in from high oil prices also funded another oil magnate that benefited from the crisis. Richard Mellon Scaife, whose family founded Gulf Oil, later became part of Chevron. In his memoir, *A Richly Conservative Life*, Scaife wrote that he was proud to have been funding contrarian climate science. Over four decades, the Scaife Foundations provided hundreds of millions to organisations such as The Heritage Foundation.[29]

Finally, the CEI is yet another conservative think tank highly influential within the climate denial machine, as shown in Figure 3.2. Created in 1984 as a libertarian think tank counting environment and energy as one of its focus themes, it has among other activities been supporting the George W. Bush administration in its efforts to oppose effective climate policies.[30] For example, the CEI released a pro-CO_2 commercial called "Energy" in 2006. The short video depicts the greatness of the progress brought about by fossil fuels and clearly draws on carbon vitalist arguments mentioned in the previous chapter when claiming that CO_2 is "essential to life". The next second it denounces "politicians who want to label carbon dioxide a pollutant – imagine if they succeed, what would our lives be like then?", playing suspenseful music in the background. It ends with a sentence that summarises their point: "Carbon dioxide: they call it pollution; we call it life".[31]

While the above think tanks have been singled out as some of the most visible and aggressive in their opposition to climate science and policy in the USA, obstructionist strategies are seen among neoliberal and conservative think tanks around the world. The influence by misinformation on science and framings of climate policies from conservative think tanks has had effects over the past two decades. Especially the organisations with ties to either the Koch brothers or Exxon have succeeded in finding the ears of politicians and been described as trustworthy in news media.[32]

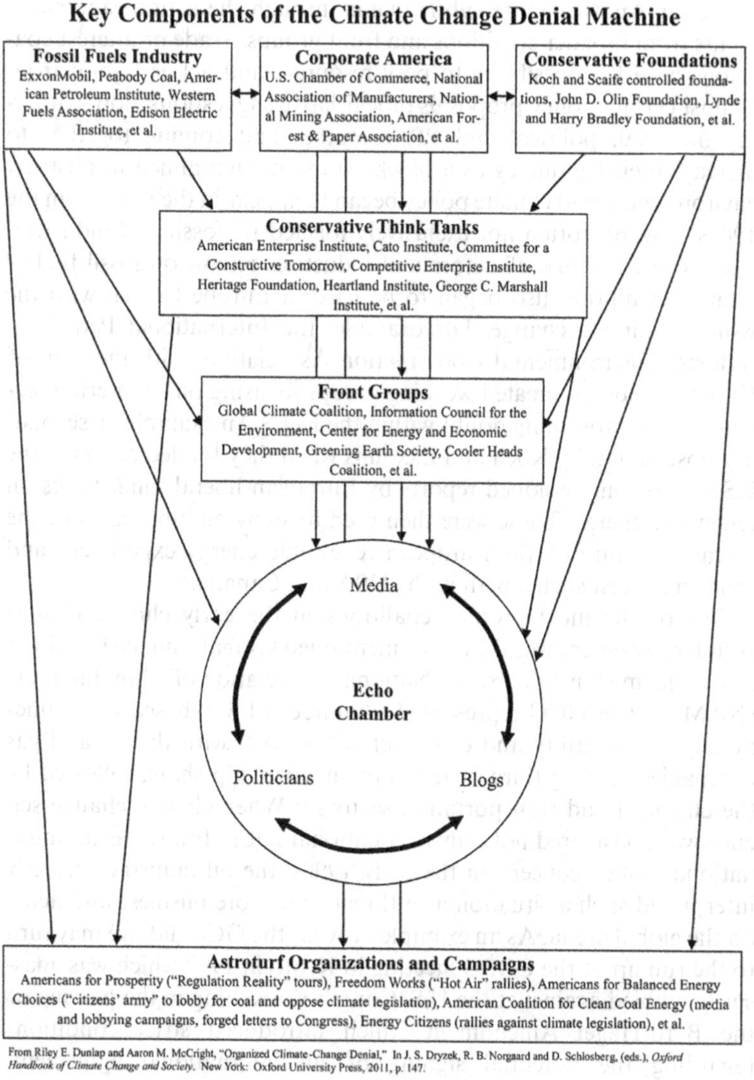

Key Components of the Climate Change Denial Machine

Fossil Fuels Industry	**Corporate America**	**Conservative Foundations**
ExxonMobil, Peabody Coal, American Petroleum Institute, Western Fuels Association, Edison Electric Institute, et al.	U.S. Chamber of Commerce, National Association of Manufacturers, National Mining Association, American Forest & Paper Association, et al.	Koch and Scaife controlled foundations, John D. Olin Foundation, Lynde and Harry Bradley Foundation, et al.

Conservative Think Tanks
American Enterprise Institute, Cato Institute, Committee for a Constructive Tomorrow, Competitive Enterprise Institute, Heritage Foundation, Heartland Institute, George C. Marshall Institute, et al.

Front Groups
Global Climate Coalition, Information Council for the Environment, Center for Energy and Economic Development, Greening Earth Society, Cooler Heads Coalition, et al.

Media

Echo Chamber

Politicians   Blogs

Astroturf Organizations and Campaigns
Americans for Prosperity ("Regulation Reality" tours), Freedom Works ("Hot Air" rallies), Americans for Balanced Energy Choices ("citizens' army" to lobby for coal and oppose climate legislation), American Coalition for Clean Coal Energy (media and lobbying campaigns, forged letters to Congress), Energy Citizens (rallies against climate legislation), et al.

From Riley E. Dunlap and Aaron M. McCright, "Organized Climate-Change Denial," In J. S. Dryzek, R. B. Norgaard and D. Schlosberg, (eds.), *Oxford Handbook of Climate Change and Society*. New York: Oxford University Press, 2011, p. 147.

Figure 3.2 The workings of the climate change denial machine.[17]

Source: reproduced from Dunlap and McCright 2011.

Front groups and coalitions

The direct funding of scientists, think tanks and lobby groups that could help spreading climate science doubt is one way to exert influence. Another more subtle one is through the actions and statements from broader coalitions and front groups. Trade or interest coalitions have the possibility to pool resources and act as a joint front against certain policy proposals or legislative suggestions, both gathering financial, political and PR resources and coming together to increase their legitimacy as a block.[33] Groups that aimed to obstruct environmental and climate policy began to appear in the USA from the 1980s onwards, often not exclusively formed by fossil fuel industries but also by a range of actors and industries reliant on fossil fuels.[34] Similar coalitions also began to be used in Europe to deal with the issue of climate change. For example, the International Petroleum Industry Environmental Conservation Association, with the help of Exxon employees, created working groups focusing on countering regulation and promoting doubt within the EU.[35] An example of secondary obstruction by Koch and Exxon is when they funded actors in the USA who commissioned reports by European liberal think tanks on renewable energy. These were then used to show multiple cases of the so-called failure of the European renewable energy experiment and caution against such a path in the USA and Canada.[36]

One of the most profiled coalitions in the early climate change countermovement was the above-mentioned Global Climate Coalition (GCC) formed in 1989 by the National Association of Manufacturers (NAM).[37] The GCC represented a number of US-based companies mainly in the utility and coal/steel/rail sectors, with the oil and gas companies coming third in terms of membership share, followed by the chemical and transportation sectors.[38] When climate change science was considered not only as a national but re-framed as an international policy concern in the early 1990s, the oil industry correctly interpreted such a situation as a threat to its core business and acted on the global arena. As an example of what the GCC did, we may turn to the run up of the COP 3 meeting in Kyoto in 1997 which was, massively lobbied against at the U.S. Congress, resulting in the passing of the Byrd-Hagel Amendment which introduced strict conditions regarding the potential signing of an international agreement. Subsequently, the CGC joined forces with other trade associations to create the Global Climate Information Project (GCIP), which initiated a $13 million PR campaign emphasising the supposed devastating economic effect of the upcoming Kyoto Protocol. This significantly influenced the public opinion before the USA eventually signed the treaty, which at that time was leaving climate budgeting with strict limits combined with planned decrease of emissions aside and shapeshifting towards carbon trade.[39]

The same year, John Browne, the Chief Executive Officer of British Petroleum (BP), made international headlines by announcing that his company was splitting from the rest of the oil industry and would support international greenhouse gas regulation. And indeed, BP (rebranding itself as "Beyond Petroleum" in 2001) and Dupont withdrew from the Global Climate Coalition. Royal Dutch Shell, Ford and then – in rapid succession in the early months of 2000 – Daimler Chrysler, Texaco, and General Motors announced that they too were leaving the coalition and instead accepted that Earth was heated due to greenhouse gas emissions mainly from burning fossil fuels. Thus, the forms of primary obstruction with literal denial of climate science by the GCC were short-lived as the overarching strategy of its members, as there were already secondary obstruction tactics in use by its company partners and tried out by oil extractive countries such as Norway via state enterprises. In fact, the above-mentioned companies claimed that they already had solutions at hand focusing on emissions such as fuel cells, electric cars, hydrogen, carbon capture and storage ignoring the problem of fossil fuel extraction. Rather unsurprisingly, their solutions came handy as they did not require these companies to change their business model drastically and can therefore be understood as secondary obstruction.[40]

However, several types of literal denial tropes of primary obstruction, pioneered by the denial machine and spread by GCC, are still in use globally.[41] They are based on pseudo-scientific arguments produced by the denial machine and spread via propaganda, which can be divided into four different groups. First, there is no ongoing heating; second, humans are not responsible; third, the consequences are positive; and fourth, if global heating is acknowledged as having negative consequences, it is nevertheless argued that these consequences are minimal in relation to other, more pressing issues.[42] Brulle emphasises how the CGC carried out four main obstruction practices: 1) monitoring and contesting climate science, 2) commissioning and utilising economic studies to amplify and legitimate their arguments, 3) shifting the cultural understanding of climate change through public relations campaigns and 4) conducting aggressive lobbying of political elites.[43] These practices have been common and still remain widely used in combination with the tropes mentioned above.

Astro-turf organisations and the mobilisation of public opinion

We have seen that think tanks are not the only structures financed by fossil fuel money, as coalitions also received funds.[44] However, money from the fossil fuel industry and conservative foundations is used in yet another way to obstruct climate change mitigation by supporting so-called astro-turf organisations.[45] Astro-turf alludes to something which appears to be real but is a fake.[46] Astro-turf campaigns are presented as emanating from grassroots movements, while they are in fact more or less unfolding top-down. Such astro-turf organisations are commonly used in environmental struggles by vested interests as a climate obstruction technique, yet they are not extensively studied.[47] More often than not, (fossil fuel) companies control the expressions from these organisations as they provide money, infrastructures and networks for the creation and communication of the ensuing (highly subsidised) campaigns.[48] Mobilising astro-turf engagement for an issue has been a PR strategy since the 1970s. During that period, PR companies recognised the power of the new social movements and their claim to speak for the public. For corporations, it was crucial to change the narrative and to make it seem like they were the ones speaking for the public interest. Countering popular protests with pseudo-public campaigns became one such strategy.[49] Examples of such campaigns are the Hot Air tours of 2008 and the Energy Citizens campaign of 2009 in the USA, two efforts created by the Koch-funded front group Americans for Prosperity and Freedom Work.[50] Another example comes from Canada, where Canada's Energy Citizens (2014) and Canada Action (2010) are two groups founded or funded by fossil fuel capital to mobilise popular support for the industry in light of climate taxes.[51] These groups were crucial in the convoy of truckers known as "United We Roll", in a less organised manner such effort to portray industry interest as the same as the public's during the winter of 2022. The energy price spikes in the period led to a sticker campaign blaming the high prices on the current leadership both in the USA and in Sweden. The campaign was originally set up by an individual affiliated to the Confederation of Swedish Enterprise in similar vein as viral memes discrediting Greta Thunberg.[52]

What astro-turf and subsidised actors for example do is to create and enhance existing conflicts in society by portraying specific policies as destructive for certain individuals. The campaigns have been efficient in maintaining conflict and enabled industry actors to draw support from such popular protests in policy negotiations. Similar campaigns can strengthen the appearance of opposition to climate action even though public opinion polls high on willingness to act on climate change.[53] The mobilisation of publics and the spread of denial in public arenas is also one key strategy that has resulted in a preserved uncertainty among the public even though science is becoming ever more certain, as we shall see in Chapter 5.

What effects by the propaganda?

The work of the Global Climate Coalition, the Heartland Institute, the George C. Marshall Institute and others in shaping climate change as an issue by propaganda paid off quickly both nationally in the U.S. as well as internationally (in addition to having long-term consequences), as they succeeded in getting scientific articles published which claimed that the science surrounding climate change threat was uncertain. The message made a splash and news media began to report on researchers who provided an alternative version to the established climate research in the early 1990s. The denial machinery had thus begun to manufacture its product: distrust of science. This led, for example, the G. W. Bush administration to change its position on the climate issue by blocking international efforts to reach a climate agreement via the U.S. delegates. At the Rio de Janeiro Environmental Summit in 1992, the USA refused to sign for the implementation of specific emission reduction targets.[54]

The influence over everyday knowledge can also be detected in voter behaviour during the years discussed here. The denial machine clearly succeeded in influencing public opinion in the USA. Several attitude surveys show that while people in Europe have become increasingly concerned about climate change since the late 1980s (with ups and downs), U.S. Americans – and Republicans in particular – are, to say the least, hesitant in acknowledging climate research. For example, between 1990 and 2010, the proportion of Republicans who trusted climate research declined.[55] In Sweden we can start observing a rise in literal denial among far-right voters after years of propaganda from far-right digital media.[56]

Political action to mitigate climate change was consequently impacted and has been slowed down at the regional, national and international levels through direct lobbying. Vested interests view climate change as a significant threat to their businesses and have been successful in making their standpoint heard, seen and influential.[57] Regions that rely on fossil fuels for jobs and revenues find it particularly difficult to withstand such lobbying attempts. Research shows how fossil fuel companies have lobbied to weaken climate policies around the world and have continued to do so while claiming to support the Paris Agreement adopted in 2015. Indeed, scholars have convincingly argued that political lobbying by fossil fuel interests is the reason why the Paris Agreement does not mention decarbonisation or the reduction of fossil fuel use, despite an overwhelming scientific agreement on the need to leave fossil fuels in the ground to limit global warming to 2°C.[58] While civil society organisations have argued that fossil fuel companies should be kept out of the UN climate change negotiations to prevent companies with a conflict of interest from participating as observers, limiting fossil fuel interests at the negotiations is complicated by the fact that state-owned companies would still have a seat at the table. Countries rich in fossil fuel reserves, such as Saudi Arabia, the USA, Kuwait and Russia, have been particularly active in obstructing the negotiations and disputing the science on climate change.[59]

Conspiracy theories and attacks on science

As conservative voters, not least in the USA, tended to dismiss climate science when entering the 21st century, an alternative worldview got hold, where results and policies connected to global warming were merged with conspiracy views. As discussed above, it is not only through pseudoscience that the denial machine is fighting the battle against climate science. Another method is to tarnish and attack climate scientists by making them appear politically biased. The early opposition to climate science was already motivated not only by economic interests but also by ideological ones. In 1988, few predicted the imminent and sudden end of the Cold War. However, the fall of the Berlin Wall and the dissolution of the Soviet Union would have profound consequences. In their book *Merchants of Doubts*, Oreskes and Conway describe how several of the early and influential actors within organised climate science denial in the USA were motivated by neoliberal ideas, seeing the USA and themselves as defenders of the free world. The USA had defeated Nazism, and after the war it stood up against totalitarian communism in Russia and China. According to such conservative actors, only the expansion of markets and free enterprise could guarantee democracy and the freedom of the individual. Representatives of the environmental movement were portrayed by opponents as watermelons, green on the outside and red on the inside. With the end of the Cold War and the fall of the Soviet Union, the balance of power which influenced the governance architecture of IPCC also shifted: the global agenda thus became more heavily dependent on the actions and position of USA. Thus, if the U.S.

climate debate could be shifted in favour of minor or no climate action, such a move would be able to change the narrative around the entire international community's response to the climate crisis. In connection with the second IPCC report stating the famous line that "the balance of evidence suggests that there is a discernible human influence on global climate", published in 1995, The Global Climate Coalition and representatives of the George C. Marshall Institute conducted a campaign to undermine the report's credibility, as noted above. The panel was accused, among other things, of deleting a sentence that emphasised the uncertainty in the research.[60] Several actors around the world, including many who are mentioned in this chapter, contested the findings of the IPCC by signing a declaration disputing the idea of scientific consensus and the need for "hasty actions".[61] The IPCC report was met with fierce resistance and in the aim of the scope came convening lead author Ben Santer. Accusations of "scientific cleansing" and unauthorised changes to parts of the report were thrown at Santer by contrarian scientists Fred Singer and Fred Seitz for the revisions that had taken place during the normal revision procedure. But despite clarifying responses from both Santer and IPCC chairman Bert Bolin the contrarians doubled down on their accusations and continued to voice them in the years to come.[62]

An effort to undermine trust in climate science on a much larger scale took place in connection to what was, by obstructionists, named "Climategate" in 2009, a few weeks before the Copenhagen climate conference COP 15. Thousands of emails were stolen from a server at the University of East Anglia (UEA) in the UK and climate scientist Philip Jones saw his sent and received emails since 1996 made available publicly. Jones was the head of the Climate Research Unit (CRU) at the UEA, who had substantially contributed to several IPCC climate reports. Research carried out over decades at the CRU had supported the case for anthropogenic climate change. The correspondence from the CRU was uploaded to various websites, one of the more influential one being *Watts Up With That*, a blog run by Anthony Watts – who was at the time and later on funded by the Heartland Institute, and even spoke at the institute's Third International Conference on Climate change earlier that summer. The term "Climategate" was first used on November 20 in a blog post by James Delingpole, a writer for the UK conservative newspaper *The Daily Telegraph* and a well-known climate obstructionist. Within the global climate denial discourse (but also beyond), the term quickly established itself, evoking memories of the Watergate scandal, and was thus designed to suggest the existence of irrefutable proof that the criticism of the IPCC had been correct all along.[63] Subsequently, the main target for the hack, Michael Mann, a professor of Earth System Science at Pennsylvania State University, was accused of fraud and lying. Mann was the main author of a historical temperature reconstruction (dating back to AD 1000 and known as the "hockey stick"), which had become highly influential among climate scientists, policymakers and the media up until Copenhagen – not least through its popularisation via the documentary *An Inconvenient Truth* based upon talks given by 2007 Nobel Peace Prize awardee Al Gore.

The illusory conspiracy came to dominate CNN's reporting during the first days of COP 15, generating considerable press attention across the USA and around the world, with articles and editorials published in major newspapers such as the *Washington Post*, scientific journals and stories broadcast on major television and radio channels.[64] Several books were also rapidly written by climate contrarians in the USA and France and used the controversy as a proof that climate change was a hoax.[65] However, charges of corruption, lies, cover-ups and fraud did not hold up to close analysis – though the perception took hold, was strategically used and hyped climate conspiracy theories.[66] This event displays how merchants of doubt fuelled a larger ideological shift as it was widely taken up by far-right political leaders and parties, which we will discuss in the next chapter.

Denial groups in Sweden, for example, started to make use of conspiracy theories as a complementary tactic which tied smoothly into ideologically far-right, party-political logics of the honest citizen against the corrupt elite.[67] In France, Sweden and the USA, the situation post-2009 has been labelled as a third phase of the climate change debate in which public controversies and contestations are part of the issue (following the first phase, of the carbon-war period of 1990–1998 dominated by fossil fuel imaginary and the formation of the climate denial machine; and the second phase, techno-market period with a carbon promise 1999–2008 of which technologies such as carbon capture and storage and fuel cells were combined into techno-capitalist ecomodern utopian visions such as the Hydrogen Economy).[68]

As an example of such historical classification, let us look at Sweden again. There, outright climate denial in the public debate was very marginal until the Stockholm Initiative (later Climate Realists) entered the scene in 2008. This network had a direct impact when targeting mass media with more than 15 letters to editors in half a year's time and also getting airtime via interviews. Up until 2012 it still had limited influence on Swedish politics.[69] However, during a parliamentary debate on January 29, 2013, the energy and environment spokesperson from the far-right party Sweden Democrats, Josef Fransson, spread climate denial tropes. Fransson claimed that the apocalyptic scenario that many painted in the climate debate was totally false. For him the climate crisis was a conspiracy made up by corrupt scientists and the lying political elite. Fransson said this was all built upon "fake climate science".[70] Literal and interpretative denial has similarly grown stronger in the USA over the decades from 1990,[71] as shown by the analysis of citizens' opinion polls, and its influence on the party-political scene can be detected in it being a major theme in Trump's successful 2016 presidential campaign, as well as by the changing of rules and regulations in favour of the fossil fuel industry during his presidency.[72] Similar trends can now be found in Europe, not least in connection with far-right political parties and right-wing nationalist digital media, which the next chapter will describe further.

A countermovement turned into a governing body

With Donald Trump's victory in the 2016 U.S. presidential election and Jair Bolsonaro elected in Brazil, the denial machine took the final big step from being a countermovement when taking control of climate policy from inside democratic institutions (authoritarian leadership in Saudi Arabia, Russia and many more we know had since long obstructing climate science and policy). As Bush had done 16 years earlier, Trump made sure the fossil fuel industry was well represented in the administration. Even though he left early on, ExxonMobil's former CEO Rex Tillerson was put in charge of foreign affairs. David Bernhard, who had worked most of his life as a lobbyist for the oil industry, was appointed Minister of the Interior, and Rick Perry, who had repeatedly expressed climate-denying arguments, as Minister of Energy.[73] At least 20 people with a background in the fossil fuel industry were put in key positions, including William Happer signatory of the Leipzig Declaration, which opposed the 1995 IPCC report, who was given a seat at the U.S. National Security Council and later became chairman of Trump's own climate panel. This same Happer had started the CO_2 Coalition that emerged from the George C. Marshall Institute funded by, among others, the Mercer family known for donating money to the Brexit campaign, Breitbart News and far-right forces in the USA. The most damaging designation was probably Scott Pruitt, who was appointed head of the environmental authority. Pruitt had previously worked closely with both the oil and coal industries and is outspoken in his distrust of climate science. Indeed, during his time as a prosecutor in Oklahoma, he tried, on a total of 14 occasions, to sue the authority for which he was later appointed head of. However, Pruitt was forced to resign after allegations of corruption and was succeeded by Andrew Wheeler, who had a similar attitude towards the climate issue. Like Pruitt, he had worked for several years as a lobbyist in the coal industry. These two men ensured that the environmental authority's budget in 2017 was cut by around 30 per cent.[74] The "swamp" of lobbyists that Trump promised to "drain" moved into the White House with him instead. It exposed how a marriage between an organised climate denial machine and a greedy misogynist showman enacting and riding a far-right populist sentiment affects the future of our planet.

Summary

From 1988 onwards, the formation of a denial machine enacting primary obstruction took place as a reaction to outer pressure and in contrast to results produced and/or acknowledged by the fossil fuel industry. To summarise, this denial machine, consisting of fossil fuel industry money, knowledge brokers, corrupt scientists and a media ecosystem, affected the public and politicians alike and has been one of the best organised and economically strongest propaganda apparatus operating in a democratic country in peacetime. It has been very successful – a reality that research particularly in the USA has painstakingly shown.[75] The USA is, besides the oil-producing countries in the Middle East, the country that most clearly opposed measures to reduce greenhouse gas emissions, and its population is the one in the Western world with the weakest trust in climate research. Over the years, the Republican Party has become a guarantor that all progressive climate policy proposals are stopped in the U.S. Congress.

There is a simple reason this lobbying machinery has been so powerful. In front of these giants and their well-greased and media-savvy PR apparatus, science ended up disadvantaged when in terms of fighting for the attention of voters and decision-makers. But this is of course not the whole reason for the success of different climate obstruction strategies. Conservative think tanks have played the role of important brokers to facilitate exchange between the primary obstruction of fossil fuel companies and the secondary obstruction of most governments around the globe. It is therefore important to notice, that the timing of our climate emergency is particularly bad considering how the Global

North nations particularly have been transformed from welfare states to market governance.[76] The primary obstruction by actors such as Exxon or the Cato Institute was even more effective, given the general shift towards the strategy of solving environmental issues with neoliberal logics. The efficiency of direct regulations was continuously downplayed by actors across the political spectrum during the 1990s were seen as the way forward, which meant that any direct interference with fossil fuel operations became increasingly hard to impose.[77]

As we have indicated towards the end of this chapter, climate obstruction has also been picked up by far-right actors, from political parties to non-parliamentary ones. These actors most clearly transcend the confines of the USA, with the key figures being not only Donald Trump but also the Brazilian Jair Bolsonaro as well as far-right parties in Germany and Sweden, for example.[78] The next chapter will consider the ideological foundation, party politics and communication of such obstruction in detail.

Notes

1 Oreskes, N. and Conway, E.M. (2010) *Merchants of Doubt: How a Handful Of Scientists Obscured the Truth on Issues from Tobacco Smoke to Global Warming*. London: Bloomsbury; Bergquist, A.-K. and Näsman, M. (2021) "Safe before green! The greening of Volvo cars in the 1970s–1990s", *Enterprise & Society*, pp.1–31. https://doi.org/10.1017/eso.2021.23.
2 International Energy Agency *World Energy Outlook 2021*.
3 Supran, G. and Oreskes, N. (2017) "Assessing ExxonMobil's climate change communications (1977–2014)", *Environmental Research Letters*, 12(8), 084019.
4 Dunlap, R.E. and McCright, A.M. (2011) "Organized climate change denial" in Dryzek, J.S., Norgaard, R.B. and Schlosberg, D. (eds) *The Oxford Handbook of Climate Change and Society*, pp.144–160. https://doi.org/10.1093/oxfordhb/9780199566600.003.0010.
5 Anker, P. (2018) "A pioneer country? A history of Norwegian climate politics", *Climatic Change*, 151(1), pp.29–41. https://doi.org/10.1007/s10584-016-1653-x; Nissen, A. (2021) "A greener shade of black? Statoil, the Norwegian government and climate change, 1990—2005", *Scandinavian Journal of History*, 46(3), pp.408–429. https://doi.org/10.1080/03468755.2021.1876757; Ytterstad, A., Houeland, C. and Jordhus-Lier, D. (2022) "Heroes of the day after tomorrow: 'the oil worker' in Norwegian climate coverage 2017–2021", *Journalism Practice*, 16(2–3), pp.317–333. https://doi.org/10.1080/17512786.2021.2002712; Norgaard, K.M. (2006) "'We don't really want to know'": environmental justice and socially organized denial of global warming in Norway", *Organization & Environment*, 19(3), pp.347–370. https://doi.org/10.1177/1086026606292571.
6 Featherstone, D. (2013) "The contested politics of climate change and the crisis of neo-liberalism", *ACME: An International E-Journal for Critical*

Geographies, 12(1), pp.44–64; Buhr, K. and Hansson, A. (2011) "Capturing the stories of corporations: a comparison of media debates on carbon capture and storage in Norway and Sweden", *Global Environmental Change*, 21(2), pp.336–345. https://doi.org/10.1016/j.gloenvcha.2011.01.021; Blühdorn, I. (2011) "The politics of unsustainability: COP15, post-ecologism, and the ecological paradox", *Organization & Environment*, 24(1), pp.34–53. https://doi.org/10.1177/1086026611402008.

7 Lamb, W.F. *et al.* (2020) "Discourses of climate delay", *Global Sustainability*, 3, e17. https://doi.org/10.1017/sus.2020.13.

8 Asdal, K. (2014) "From climate issue to oil issue: offices of public admin-istration, versions of economics, and the ordinary technologies of poli-tics", *Environment and Planning A*, 46(9), pp.2110–2124. https://doi.org/10.1068/a140048p; Lahn, B. and Rowe, E.W. (2014) "How to be a 'front-runner': Norway and international climate politics" in De Carvalho, B. and Neumann, I.B. (eds) *Small State Status Seeking: Norway's Quest for International Standing*. Abingdon: Routledge.

9 Stoddard, I. *et al.* (2021) "Three decades of climate mitigation: why hav-en't we bent the global emissions curve?", *Annual Review of Environment and Resources*, 46, pp.653–689. https://doi.org/10.1146/annurev-environ-012220-011104.

10 McCright, A.M. and Dunlap, R.E. (2003) "Defeating Kyoto: the conserv-ative movement's impact on U.S. climate change policy", *Social Problems*, 50(3), pp.348–373. https://doi.org/10.1525/sp.2003.50.3.348.

11 Cann, W.H. and Raymond, L. (2018) "Does climate denialism still matter? The prevalence of alternative frames in opposition to climate policy", *Environmental Politics*, 27(3), pp.433–454. https://doi.org/10.1080/096440 16.2018.1439353.

12 Jastrow, R., Nierenberg, W.A. and Seitz, F. (1991) "Global warming: what does the science tell us?", *Energy*, 16(11–12), pp.1331–1345; see also Lahsen, M. (2008) "Experiences of modernity in the greenhouse: a cul-tural analysis of a physicist 'trio' supporting the backlash against global warming", *Global Environmental Change*, 18(1), pp.204–219; Oreskes, N. and Conway, E.M. (2010) "Defeating the merchants of doubt", *Nature*, 465(7299), pp.686–687. https://doi.org/10.1038/465686a.

13 Cook, J. *et al.* (2019) "America misled: how the fossil fuel industry deliber-ately misled Americans about climate change", Fairfax, VA: George Mason University Center for Climate Change Communication. Available at: https://www.climatechangecommunication.org/america-misled/ [Last accessed March 29, 2022].

14 Ibid., p.7.

15 Brulle, R.J. (2022) "Advocating inaction: a historical analysis of the Global Climate Coalition", *Environmental Politics*. https://doi.org/10.1080/09644016.2022.2058815.

16 Newell, P. and Paterson, M. (1998) "A climate for business: global warm-ing, the state and capital", *Review of International Political Economy*, 5(4), pp.679–703.

17 Dunlap, R.E. and McCright, A.M. (2011) "Organized climate change denial", p.147.

18 Brulle, R.J. (2014) "Institutionalizing delay: foundation funding and the creation of U.S. climate change counter-movement organizations",

Climatic Change, 122, pp.681–694. https://doi.org/10.1007/s10584-013-1018-7.
19 Brulle, R.J. (2018) "The climate lobby: a sectoral analysis of lobbying spending on climate change in the USA, 2000 to 2016", *Climatic Change*, 149, pp.289–303. https://doi.org/10.1007/s10584-018-2241-z.
20 Brulle, R.J. (2014) "Institutionalizing delay"; Brulle, R.J. (2018) "The climate lobby".
21 Plehwe, D. (2021) "Think tanks and the politics of climate change" in Abelson, D.E. and Rastrick, C.J. (eds) *Handbook on Think Tanks in Public Policy*. Cheltenham: Edward Elgar Publishing, pp.150–165. https://doi.org/10.4337/9781789901849; Plehwe, D. (2022) "Reluctant transformers or reconsidering opposition to climate change mitigation? German think tanks between environmentalism and neoliberalism", *Globalizations*. https://doi.org/10.1080/14747731.2022.2038358; Moreno, J.A., Kinn, M. and Narberhaus, M. (2022) "A stronghold of climate change denialism in Germany: case study of the output and press representation of the Think Tank EIKE", *International Journal of Communication*, 16, pp.267–288.
22 Donors Trust, "Mission & Principles". Available at: https://www.donorstrust.org/who-we-are/mission-principles/ [Last accessed March 29, 2022].
23 Brulle, R.J. (2014) "Institutionalizing delay"; Aronczyk, M. and Espinoza, M.I. (2022) *A Strategic Nature: Public Relations and the Politics of American Environmentalism*. Oxford University Press; Brulle, R.J., Aronczyk, M. and Carmichael, J. (2020) "Corporate promotion and climate change: an analysis of key variables affecting advertising spending by major oil corporations, 1986-2015", *Climatic Change*, 159, pp.87–101. https://doi.org/10.1007/s10584-019-02582-8; Brulle, R.J. and Werthman, C. (2021) "The role of public relations firms in climate change politics", *Climatic Change*, 169(8). https://doi.org/10.1007/s10584-021-03244-4.
24 DeSmog. "Atlas Network (Atlas Economic Research Foundation)". Available at: https://www.desmog.com/atlas-economic-research-foundation/ [Last accessed April 21, 2022]; Mckie, R.E. *et al.* (2021) "The mises institute network and climate policy. 9 findings", *CSSN Research Report*, 2.
25 McKie, R. (2022) "Climate change counter movement organisations: an international deviant network?" in Tindall, D., Stoddart, M.J. and Dunlap, R. (eds) *Handbook of Anti-Environmentalism*. Edward Elgar. https://doi.org/10.4337/9781839100222.
26 McKewon, E. (2012) "TALKING POINTS AMMO: the use of neoliberal think tank fantasy themes to delegitimise scientific knowledge of climate change in Australian newspapers", *Journalism Studies*, 13(2), pp.277–297. https://doi.org/10.1080/1461670X.2011.646403; Farrell, J. (2016) "Network structure and influence of the climate change counter-movement", *Nature Climate Change*, 6(4), pp.370–374. https://doi.org/10.1038/nclimate2875; Hudson, M. (2020) "Enacted inertia: Australian fossil fuel incumbents' strategies to undermine challengers" in Wood, G. and Baker, K. (eds) *The Palgrave Handbook of Managing Fossil Fuels and Energy Transitions*. Palgrave Macmillan, pp.195–222. https://doi.org/10.1007/978-3-030-28076-5_8; Almiron, N. and Xifra, J. (2019) *Climate Change Denial and Public Relations: Strategic Communication and Interest Groups in Climate Inaction*. Abingdon, Oxon: Routledge; Young, N. and Coutinho, A. (2013) "Government, anti-reflexivity, and

the construction of public ignorance about climate change: Australia and Canada compared", *Global Environmental Politics*, 13(2), pp.89–108. https://doi.org/10.1162/GLEP_a_00168; Ruser, A. (2021) "Widening the gap: US think tanks and the manufactured chasm between scientific expertise and common sense on climate change", Landry, J. (ed) *Critical Perspectives on Think Tanks*. Edward Elgar Publishing. https://doi.org/10.4337/9781789909234.

27 Boussalis, C. and Coan, T.G. (2016) "Text-mining the signals of climate change doubt", *Global Environmental Change*, 36, pp.89–100. https://doi.org/10.1016/j.gloenvcha.2015.12.001.

28 Idso, C.D. (2014) "S. Fred Singer and the nongovernmental international panel on climate change", *Energy & Environment*, 25(6–7), pp.1137–1148.

29 Dunlap, R.E. and McCright, A.M. (2010) "Climate change denial: sources, actors and strategies" in Lever-Tracy, C. (ed) *Routledge Handbook of Climate Change and Society*. New York: Routledge, pp.270–290; Brulle, R.J. (2021) "Networks of opposition: a structural analysis of U.S. climate change countermovement coalitions 1989–2015", *Sociological Inquiry*, 91, pp.603–624. https://doi.org/10.1111/soin.12333; Farrell, J. (2016) "Corporate funding and ideological polarization about climate change", *Proceedings of the National Academy of Sciences*, 113(1), pp.92–97. https://doi.org/10.1073/pnas.1509433112.

30 Dunlap, R.E. and McCright, A.M. (2011) "Organized climate change denial", p.149.

31 Competitive Enterprise Institute (2006) "Energy", *YouTube*, 18 May. Available at: https://www.youtube.com/watch?v=7sGKvDNdJNA [Last accessed March 30, 2022]. For a post-colonial analysis of this video, see Malm, A. and the Zetkin Collective (2021) *White Skin, Black Fuel: On the Danger of Fossil Fascism*. London: Verso, pp.19–21.

32 Farrell, J. (2016) "Network structure and influence of the climate change counter-movement".

33 Brulle, R.J. (2022) "Advocating inaction".

34 Brulle, R.J. (2021) "Networks of opposition".

35 Bonneuil, C., Choquet, P.L. and Franta, B. (2021) "Early warnings and emerging accountability: total's responses to global warming, 1971–2021", *Global Environmental Change*, 71, 102386. https://doi.org/10.1016/j.gloenvcha.2021.102386.

36 Peter, H. (2011) "Gray Studies vs. Green Jobs", Think Tank Network Research Initiative, 29 August. Available at: http://thinktanknetwork research.net/blog_ttni_en/gray-studies-vs-green-jobs/ [Last accessed March 29, 2022].

37 Franz, W. (1998) "Science, skeptics and non-state actors in the greenhouse", *ENRP Working Paper E-98-18*, Harvard University. Retrieved in Brulle, R.J. (2022) "Advocating inaction: a historical analysis of the Global Climate Coalition".

38 Brulle, R.J. (2022) "Advocating inaction".

39 Ibid; Castree, N. (2008) "Neoliberalising nature: the logics of deregulation and reregulation", *Environment and Planning A*, 40(1), pp.131–152. https://doi.org/10.1068/a3999.

40 Levy, D.L. and Kolk, A. (2002) "Strategic responses to global climate change: conflicting pressures on multinationals in the oil industry", *Business and Politics*, 4(3), pp.275–300. https://doi.org/10.2202/1469-3569.1042; Hultman, M. and Nordlund, C. (2013) "Energizing technology: expectations of fuel cells and the hydrogen economy, 1990–2005", *History and Technology*, 29(1), pp.33–53. https://doi.org/10.1080/0734151 2.2013.778145.

41 Boussalis, C. and Coan, T.G. (2016) "Text-mining the signals of climate change doubt".

42 Dunlap, R.E and McCright, A.M. (2015) "Challenging climate change: the denial countermovement" in Dunlap, R.E. and Brulle, R.J. (eds) *Climate Change and Society: Sociological Perspectives*. Oxford University Press, pp.300–332. https://doi.org/10.1093/acprof:oso/9780199356102.003. 0010.

43 Brulle, R.J. (2022) "Advocating inaction".

44 Dunlap, R.E. and McCright, A.M. (2015) "Challenging climate change".

45 Kraemer, R., Whiteman, G. and Banerjee, B. (2013) "Conflict and astroturfing in Niyamgiri: the importance of national advocacy networks in anti-corporate social movements", *Organization Studies*, 34(5–6), pp.823–852. https://doi.org/10.1177/0170840613479240; Hobbs, M., Della Bosca, H., Schlosberg, D. and Sun, C. (2020) "Turf wars: using social media network analysis to examine the suspected astroturfing campaign for the Adani Carmichael Coal mine on Twitter", *Journal of Public Affairs*, 20(2), e2057. https://doi.org/10.1002/pa.2057; Bsumek, P.K., Schneider, J., Schwarze, S. and Peeples, J. (2014) "Corporate ventriloquism: corporate advocacy, the coal industry, and the appropriation of voice" in Peeples, J. and Depoe, S. (eds) *Voice and Environmental Communication*. London: Palgrave Macmillan, pp.21–43. https://doi.org/10.1057/9781137433749_2.

46 Boykoff, M.T. and Olson, S.K. (2013) "'Wise contrarians': a keystone species in contemporary climate science, politics and policy", *Celebrity Studies*, 4(3), pp.276–291. https://doi.org/10.1080/19392397.2013.831618; Carroll, W. *et al.* (2018) "The corporate elite and the architecture of climate change denial: a network analysis of carbon capital's reach into civil society", *Canadian Review of Sociology/Revue canadienne de sociologie*, 55(3), pp.425–450. https://doi.org/10.1111/cars.12211.

47 Aronczyk, M. and Espinoza, M.I. (2021). *A Strategic Nature*.

48 Bell, S.E., Fitzgerald, J. and York, R. (2019) "Protecting the power to pollute: identity co-optation, gender, and the public relations strategies of fossil fuel industries in the United States", *Environmental Sociology*, 5(3), pp.323–338. https://doi.org/10.1080/23251042.2019.1624001; Grasso, M. (2020) "Towards a broader climate ethics: confronting the oil industry with morally relevant facts", *Energy Research and Social Science*, 62, 101383. https://doi.org/10.1016/j.erss.2019.101383.

49 Walker, E.T. (2014) *Grassroots for Hire: Public Affairs Consultants in American Democracy*. Cambridge: Cambridge University Press. https://doi.org/10.1017/CBO9781139108829.

50 Dunlap, R. E. and Brulle, R. J. (2020) "Sources and amplifiers of climate change denial". In L. M. D. C. Holmes and L. M. Richardson (eds)

Research Handbook on Communicating Climate Change. Cheltenham: Edward Elgar, pp. 49-61.

51 McLean, J. (forthcoming 2023) "United they roll? How Canadian fossil capital subsidizes the far-right" in Allen, I., Ekberg, K., Holgersen, S. and Malm, A. (eds) *Fanning the Flames: Political Ecologies of the Far Right*. Manchester: Manchester University Press; Gunster, S., Neubauer, R., Bermingham, J. and Massie, A. (2021) "'Our oil': Extractive populism in Canadian social media" in Carroll, W.K. (ed) *Regime of Obstruction: How Corporate Power Blocks Energy Democracy*. AU Press. https://doi.org/10.15215/aupress/9781771992893.01.

52 Keller, J. (2021). "'This is oil country': mediated transnational girlhood, Greta Thunberg, and patriarchal petrocultures", *Feminist Media Studies*, 21(4), pp.682–686, https://doi.org/10.1080/14680777.2021.1919729; White, M. (2022) "Greta Thunberg is 'giving a face' to climate activism: confronting anti-feminist, anti-environmentalist, and ableist memes", *Australian Feminist Studies*, https://doi.org/10.1080/08164649.2022.2062667.

53 Aronczyk, M. and Espinoza, M.I. (2022) *A Strategic Nature*, pp.96–126.

54 Parson, E.A., Haas, P.M. and Levy, M.A. (1992) "A summary of the major documents signed at the Earth Summit and the Global Forum", *Environment: Science and Policy for Sustainable Development*, 34(8), pp.12–36.

55 Weber, E.U. and Stern, P.C. (2011) "Public understanding of climate change in the United States", *American Psychologist*, 66(4), pp.315–328. https://doi.org/10.1037/a0023253.

56 Vowles, K. and Hultman, M. (2021) "Scare-quoting climate: the rapid rise of climate denial in the Swedish far-right media ecosystem", *Nordic Journal of Media Studies*, 3(1), pp.79–95. https://doi.org/10.2478/njms-2021-0005

57 Nasiritousi, N. (2017) "Fossil fuel emitters and climate change: unpacking the governance activities of large oil and gas companies", *Environmental Politics*, 26(4), pp.621–647. https://doi.org/10.1080/09644016.2017.1320832.

58 Piggot, G., Erickson, P., van Asselt, H. and Lazarus, M. (2018) "Swimming upstream: addressing fossil fuel supply under the UNFCCC", *Climate Policy*, 18(9), pp.1189–1202. https://doi.org/10.1080/14693062.2018.1494535.

59 Cadman, T., Radunsky, K., Simonelli, A. and Maraseni, T. (2018) "From Paris to Poland: a postmortem of the climate change negotiations", *International Journal of Social Quality*, 8(2), pp.27–46.

60 Oreskes, N. and Conway, E.M. (2010) *Merchants of Doubt*, p.207.

61 See the Leipzig Declaration on Global Climate change, following the International Symposium on the Greenhouse Controversy, held in Leipzig, Germany on November 9–10, 1995. Available at: https://web.archive.org/web/19981206090752/; http://www.sepp.org/leipzig.html [Last accessed March 29, 2022].

62 For the full development of this conflict see Oreskes, N. and Conway, E.M. (2010) *Merchants of Doubt*, pp.197–215.

63 Lahsen, M. (2013) "Anatomy of dissent: a cultural analysis of climate skepticism", *American Behavioral Scientist*, 57(6), pp.732–753. https://doi.org/10.1177/0002764212469799; Leiserowitz, A.A. *et al.* (2013) "Climategate, public opinion, and the loss of trust", *American Behavioral Scientist*, 57(6), pp.818–837. https://doi.org/10.1177/0002764212458272;

Nerlich, B. (2010) "'Climategate': paradoxical metaphors and political paralysis", *Environmental Values*, 19(4), pp.419–442; Grundmann, R. (2013) "'Climategate' and the scientific ethos", *Science, Technology, & Human Values*,38(1),pp.67–93.https://doi.org/10.1177/0162243911432318.
64 Raman, S. and Pearce, W. (2020) "Learning the lessons of Climategate: a cosmopolitan moment in the public life of climate science", *Wiley Interdisciplinary Reviews: Climate Change*, 11(6), e672. https://doi.org/10. 1002/wcc.672.
65 Leiserowitz, A.A. *et al.* (2013) "Climategate, public opinion, and the loss of trust"; Aykut, S.C., Comby, J.B. and Guillemot, H. (2012) "Climate change controversies in French mass media 1990–2010", *Journalism Studies*,13(2),pp.157–174.https://doi.org/10.1080/1461670X.2011.646395; Oreskes, N. and Conway, E.M. (2010) *Merchants of Doubt*.
66 Lahsen, M. (2013) "Anatomy of dissent"; Leiserowitz, A.A. *et al.* (2013) "Climategate, public opinion, and the loss of trust"; Nerlich, B. (2010) "'Climategate'"; Grundmann, R. (2013) "'Climategate' and the scientific ethos".
67 Anshelm, J. and Hultman, M. (2014) "A green fatwā? Climate change as a threat to the masculinity of industrial modernity", *NORMA*, 9(2), pp.84–96. https://doi.org/10.1080/18902138.2014.908627.
68 Aykut, S.C., Comby, J.B. and Guillemot, H. (2012) "Climate change controversies in French mass media 1990–2010"; Levy, D.L. and Spicer, A. (2013) "Contested imaginaries and the cultural political economy of climate change", *Organization*, 20(5), pp.659–678. https://doi.org/10.1177/1350508413489816; Anshelm, J. and Hultman, M. (2014) "A green fatwā?".
69 Anshelm, J. and Hultman, M. (2014) *Discourses of Global Climate Change: Apocalyptic Framing and Political Antagonisms*. Abingdon, Oxon: Routledge.
70 Hultman, M. (2020) "El viaje de la derecha nacionalista al ecocidio. El caso de los Demócratas de Suecia (Sverigedemokraterna)", *Ecología Política*, (59), pp.101–106.
71 Dunlap, R.E. and McCright, A.M. (2015) "Challenging climate change".
72 Dunlap, R.E., McCright, A. and Yarosh, J.H. (2016) "The political divide on climate change: Partisan polarization widens in the US", *Environment: Science and Policy for Sustainable Development*, 58(5), pp.4–23. https:// doi.org/10.1080/00139157.2016.1208995; Cook, J. *et al.* (2019) "America misled"; Brulle, R.J., Aronczyk, M. and Carmichael, J. (2020) "Corporate promotion and climate change"; Brulle, R.J. (2020) "Denialism: organized opposition to climate change action in the United States" in Konisky, D.M. and O'Neill, P.H. (eds) *Handbook of US Environmental Policy*. E-book: Edward Elgar Publishing.
73 Pasek, A. (2021) "Carbon vitalism: life and the body in climate denial", *Environmental Humanities*,13(1),pp.1–20.https://doi.org/10.1215/22011919-8867175.
74 Bomberg, E. (2021) "The environmental legacy of President Trump", *Policy Studies*, 42(5–6), pp.628–645. https://doi.org/10.1080/01442872.202 1.1922660; Farrell, J., McConnell, K. and Brulle, R.J. (2019) "Evidence-based strategies to combat scientific misinformation", *Nature Climate Change*, 9, pp.191–195. https://doi.org/10.1038/s41558-018-0368-6.

75 Reseach on other regions and nations are slowly catching up, showing similar results. Read for example: Bonneuil, C., Choquet, P.L. and Franta, B. (2021) "Early warnings and emerging accountability".
76 Abboud, A., Walker, J. and Mirowski, P. (2013) "Beyond denial: neoliberalism, climate change and the left", *Overland*, (210), pp.80–86. http://hdl. handle.net/10453/30883
77 Ekberg, K. and Pressfeldt, V. (forthcoming 2022) "A road to denial: climate change and neoliberal thought in Sweden, 1988–2000", *Contemporary European History*.
78 Malm, A. and the Zetkin Collective (2021) *White Skin, Black Fuel.*

4 The far right and climate obstruction

Introduction

As we have seen in the previous two chapters, organised obstruction by the corporate sector has a long history – but so have responses by activists and researchers who have stepped up to the challenge by shedding light on those driving the "denial machine". Yet, actors behind the latter are not the only ones obstructing climate action. Rather, and increasingly so since the financial crisis of the late 2000s, the climate crisis has been accompanied by a political crisis in many (Western) countries: a crisis of trust in political elites and the promise of liberal democracy to deliver for "the people". It is this crisis which is both a cause of and perpetuated by the rise of a key political force of our times, the far right – a force which has also joined debates over climate change.

Its globally best-known incarnation is former U.S. president Donald J. Trump (2017–2021), with his administration, as discussed in the previous chapter, being intertwined with fossil fuel interest groups and members of the denial machine. Unsurprisingly, thus, the USA under Trump withdrew from the Paris Agreement.[1] Resembling Trump, Jair Bolsonaro's successful campaign for Brazil president (2019–) has resulted in the dismantling of institutions charged with the protection of the environment as the latter were made responsible for "holding back" the country's economy, following years of reducing the rate of deforestation and engaging in global efforts to combat climate change.[2] However, the far right has in fact long been at home in many countries across the world, first and foremost in Europe. As such, the European experience, which has been most extensively studied, should be of interest to others – though scholarship on Europe must also learn from cases around the world.

DOI: 10.4324/9781003181132-4

But what is the far right exactly and why should we bother to consider such actors in this (unfortunately not so short) story about climate obstruction? Let us briefly clarify these two questions – "What is the far right?" and "Why should we bother?" – before looking in greater detail at how the far right has been engaged in climate obstruction.

What we call the far right today can be traced back more than two hundred years – and although it has shared ideas with conservatives and (non-far-right) nationalists, its unique selling point is, first, the essentialisation of (what is actually socially constructed) inequality. That is, the far-right views differences between, for example, ethnic, national and racial groups as well as men and women as "natural". With regard to the ethnic, national and/or racial "nature" of group members, some view these groups as being hierarchically ordered (e.g., white people being superior to non-white people), while others avoid such judgement by speaking of a "right to difference" and arguing for segregation of cultures.[3] In both cases, the far right is exclusionist, viewing non-native elements as a threat to the ideal of an ethnically homogeneous community, for example, a nation state. Connected to such ethnonationalism is, second, a sharp authoritarianism which spans from social-authoritarian values and beliefs to a preference for law-and-order policies and, in its extreme form, for a *Führerstaat* in which all authority lies, ultimately, with the leader. Of course, not every far-right actor wishes for the latter – there are vast differences between, for example, a politician of the Danish People's Party who agitates against (non-European) immigration (within the parliamentary system), a magazine or blog which attempts to shift public understandings towards far-right ideas and terrorists such as Brenton Tarrant, who killed 51 people in 2019 in Christchurch, New Zealand. Importantly, this range illustrates that the far right is not limited to party politics but that there is a multifaceted continuum of far-right actors. Furthermore, while this continuum is united in its hostility to immigrant groups viewed as "polluting" the community, it has been conceptualised as ranging from the so-called radical right (which is opposing *liberal* democracy) to the extreme right, which opposes democracy itself and commonly resorts to violent means. Separating this political spectrum from the conventional (conservative) right is not always easy as personal, organisational and/or ideological overlaps might exist. Demarcating the far right from right-wing, conservative actors has been further complicated in the course of what is called the fourth wave of post-war far-right politics since the beginning of the 21st century as "mainstream" parties and media have adopted ideas and language of the far right.[4] However, given our aim in *Climate Obstruction* and our discussion in the previous chapter, we will limit ourselves to actors which are commonly accepted as being "far-right".[5]

But why should we consider this political spectrum in the first place, given that multinational corporations, with their vast (financial) resources, affect climate policy in various, powerful ways while, to date, only a few far-right parties have dominated national governments? Furthermore, the need to consider far-right positions on climate change cannot simply derive from cases such as that of terrorists like Tarrant: although the latter addressed climate change when justifying the mass shooting – speaking of "overpopulation" engendered by non-white "invaders", which allegedly needed to be addressed to save the environment – he and his kind of eco-fascists remain, in 2022, on the fringe. However, and beyond their direct impact through, for example, legislation, there are at least two reasons which make the far right a relevant object when thinking about climate obstruction. First, some of its ideas and interpretations, such as viewing overpopulation as *the* problem instead of unequal distribution of emissions (between the Global North and South, but also within the North) and further arguments we discuss below, are often uncritically shared amongst the wider population or only become points of contestation because of the far right's agitation. Such ideas and interpretations need to be identified and called out. Second, and as we have mentioned above, because far-right actors utilise partly different arguments from those we have encountered in previous chapters, a focus on their stances provides a more comprehensive understanding of contemporary obstruction. This is especially important as far-right parties (but also non-party actors) and their political mobilisation have become more central around the world. Indeed, at the heart of this chapter, and complementing the focus of the previous two, is an interest in how far-right ideology is intertwined with climate obstruction. That is, we discuss why various forms of obstruction have been relatively widespread in the far right and, in doing so, introduce their obstructionist views.

When it comes to the far right, it has been argued that explanations which simply refer to economic and political marginalisation (those "left behind") by societal change, maybe most famously globalisation, fall short and should be complemented by an understanding of the role of ideas (and strategic communication of such ideas).[6] First and foremost, such ideas concern an aversion towards what is global, general and abstract – something certainly true for climate change which knows no borders and cannot be adequately tackled from within a bounded space.[7] In other words, as the far right does not commonly view climate change in relation to a particular, territorial setting, for example: a national homeland (even though it will certainly affect this homeland) but instead associates climate change with a global, cosmopolitan outlook (and the elite perpetuating such an agenda), much of what is associated with climate protection is not "in tune with" far-right *nationalists*.[8] Further elements in this jigsaw of ideas are, for example, concerns about the loss of national sovereignty (considering global responses to climate change) and anxieties rooted in threatened, male gender roles as industrial jobs and energy supplies change. Indeed, a 2019 review notes that "many, though not all, far-right party and non-party actors are sceptical towards (anthropogenic) climate change and/or responses to it".[9] That is, some of them still deny anthropogenic climate change and thus engage in what we call primary obstruction while others limit themselves to secondary obstruction by casting doubt on the legitimacy of decision-making processes and policy responses.[10] Clearly, whether such ideas become efficacious also depends on the so-called political opportunity structures, that is, settings, such as the party system in a given state, in which political actors operate.[11] Concerning climate change in particular, the fact that climate politics and policy have more and more become associated with the political left, and given that even the centre-right has largely accepted the need for climate mitigation (however insufficient), a space for various types of climate obstruction is vacant in the political arena or created as a political option by these actors. Hence, the focus in this chapter is on claims-making and framing of climate change by the far right: on how these actors define climate change (policy) as problematic; on what causal interpretation they offer; on how they evaluate this; and on what solutions they propose. In short: what is the audience told to believe?[12]

Having set the scene, let us turn to how the far right articulates climate obstruction. We will start by encountering primary obstruction, what has also been called "evidence scepticism" or, simply, outright denial. As we will see, this often straightforwardly conspirational view on climate change at times lives up to the worst caricatures imaginable. Subsequently, we turn to what we call secondary obstruction, that is, positions which do not any longer bother to question the science behind climate change but which nevertheless obstruct efforts and articulate subjectivities that rage against climate protection, tactically or out of conviction. Finally, we briefly address how such obstructionist communication has manifested itself in voting behaviour and in governing climate change.

"A convenient scam": why some in the far right deny the scientific evidence for anthropogenic climate change

Once upon a time, in December 2009 to be precise, world leaders met in Copenhagen to discuss climate change at the 15th Session of the Conference of the Parties to the United Nations Framework (COP15). Although hopes were high, disappointment ultimately prevailed as no binding agreement was achieved. Amongst those who had campaigned against such an agreement – though hardly responsible for the outcome – was an extreme-right British party, the British National Party (BNP).[13] Back then, the BNP was in ascend and considerable media attention had built up as the party tried to present itself as a reformed, serious far-right force under the leadership of the Cambridge University-educated Nick Griffin (in fact, the BNP remained little more than a neo-Nazi group). However, this party which had just won two seats in the European Parliament dedicated considerable resources to a campaign tasked to reveal what they perceived to be "a convenient scam".[14]

In fact, the BNP had put together a briefing paper entitled "Debunking Global Warming" to mark COP15, which attempted to refute the scientific evidence for anthropogenic climate change. Amongst much "evidence" were claims that the climate has always changed (irrespective of anthropogenic CO_2 emissions), illustrated by the statements that there was once farming in Greenland (a classic within obstructionist argumentation), that extreme weather events have not increased in numbers and that sunspots have a direct effect on warm weather. Griffin would go on and spread his views while being a Member of the European Parliament (2009–2014), speaking of "a Josef Goebbels-scale big lie from the green industrial complex" to benefit, for example, banks at the expense of "the little people".[15] And while this episode of far-right denialism is long past, Griffin occasionally returns to the subject on social media and even shared a denialist newspaper article on the eve of COP26.[16]

Another reason to start with this episode is that while we lack substantial longitudinal analyses of far-right stances on climate change, the second half of the 2000s seems to be a key moment in which arguments from the denial machine were manifestly taken up by such actors. For example, the so-called Climategate in the context of COP15 concerned a dossier of thousands of hacked emails from climate scientists which, according to denialists, "proved" that scientists manipulated data. While this has long been refuted, it was also widely received by the far right.[17] Similarly, the Danish People's Party organised an alternative to COP15 where, among others, Fred Singer, lead author of Heartland Institute's NIPCC report (which we previously encountered in Chapter 3), spoke.[18]

While the Heartland Institute will be mentioned again in this chapter, let us return to the extreme-right end of the far-right continuum. For example, Walter Lüftl, an Austrian engineer who, in the early 1990s, questioned the technical possibility of National Socialist mass murder through gas, later established a direct line from his Holocaust denialism to the denial of climate change, all under the umbrella of the "lies of our times".[19] While these "lies" differ in their content, they are all allegedly perpetuated to hold down "the people". The allure of such conspiracy theories lies in their promise to reveal what others cannot see, and to do so not only alone but as part of a community of illuminated. Indeed, only rarely is "knowledge" and "identity" so manifestly intertwined as in groups held together by beliefs in a given conspiracy. Such theories offer "unsubstantiated explanations of events or circumstances that accuse powerful malevolent groups of plotting in secret for their own benefit against the common good".[20] Of course, conspiracy beliefs are not limited to the far right, though far-right epistemic communities draw on a history of conspiratorial beliefs which can be connected to climate change, such as that of the "Jewish conspiracy". Examples pertaining to the case of climate change include beliefs in scientists faking data and "green industries"/individuals, such as the former U.S. vice president and climate activist Al Gore, furthering a climate agenda for their own benefits. Indeed, while ideology predicts stances towards climate change, conspiracy thinking does so too: more specifically, work on the USA and the Netherlands has not only argued that conspiracy theories are more prevalent amongst the extreme left and right (both prone towards conspiracy thinking) but that the latter is especially likely to believe in conspiracies related to science, including climate change.[21] Moreover, and as we will discuss below in greater detail, the authors point to the significance of gender, stating that "climate conspiracy theories flourish particularly among right-wing extremist men".[22] Overall, in the context of wider post-truth politics,[23] conspiracy beliefs have circulated widely as traditional media and epistemic authorities have partly lost their gatekeeping function and are not trusted amongst some demographics.

While it is easy to chuckle about such conspiracy beliefs and while we should not overestimate them, they hold at least two lessons: first, parts of the far right (and societies in general) live in denial land. This is interesting to the extent that others belonging to the far right accept the fact that humans are primarily responsible for climate change – though such acceptance commonly (though not exclusively) remains weeded to opposition to mainstream climate policies. Second, the primary obstruction encountered so far is not limited to the extreme right – though not many others might adopt their overly conspirational tone.

This is where we move to the present and consider the classic climate-obstructionist, far-right party in Europe of our days: the Alternative for Germany (AfD). The party's stance is enshrined in its party programme, which states that the "climate has changed since the earth came to exist", and questions the influence of human-induced CO_2 emissions.[24] However, the party's position was not always as straightforward: shortly after it was founded in early 2013, in its first election manifesto for the general election later that year, the AfD did not discuss anthropogenic climate change – though already the manifesto for the 2014 European elections spoke of alleged uncertainty regarding the causes of climate change. The party manifesto for the most recent general election in 2021 calls for adaptation and against a "hopeless fight against the changing climate", claiming that it "has not yet been proven that humans, especially industry, are significantly responsible for the change in the climate".[25] Indeed, since 2016, such primary obstruction – paired with secondary obstruction we will turn to shortly – has characterised climate communication by AfD representatives. This includes the organisation of an Alternative Climate Symposium in 2019 inside the German parliament, featuring the party's environmental spokesperson, Karsten Hilse, and international speakers, such as the British climate change denier Lord Christopher Monckton.[26]

However, it would be mistaken to believe that the AfD has been the only party organising such events during the 2010s: in 2016, the far-right Swedish Democrats organised an alternative Nobel party in Stockholm. One of the almost 400 guests from several European right-wing parties was former Czech President Václav Klaus who had risen to fame amongst climate obstructionists via his 2007 book *Blue Planet in Green Shackles: What Is Endangered: Climate or Freedom?*. Defending the nation state against what they see as illegitimate interference by the European Union, the organisers honoured Klaus with the European Freedom Prize. Having long stressed "freedom" (against his experience of communism), the neoliberal Klaus has indeed been a loud voice against climate politics which he views as a threatening doctrine or religion. Also honoured was Nigel Farage whose United Kingdom

Independence Party (UKIP) was not only instrumental in the United Kingdom (UK) exiting the European Union, as well as in mainstreaming far-right policies in the UK but which was also the primary voice of (primary) obstruction in the country.[27] Initially a populist, Eurosceptic party, Farage and UKIP moved towards the far right since early 2010,[28] with UKIP becoming the biggest British party in the European Parliament in 2014. Amongst those MEPs, Julia Reid not only denied anthropogenic climate change but claimed, like the carbon vitalists we spoke of in Chapter 2, that "CO_2 is not a pollutant".[29] Other political actors who have spread primary obstructionism include the "(grand) father" of the French far right, Jean-Marie Le Pen, as well as the Dutch anti-Islam radical-right Party for Freedom and the radical-right Spanish party Vox.[30]

Moreover, primary obstruction is not limited to political parties. Research on Austrian, German and Swedish far-right non-party sources, online and offline, has shown that these too circulate denial and beyond. Relatedly, the German far-right magazine *Compact*, which had published a strongly outright denialist special issue in 2017, organised a "Conference against the Climate Mania" in 2019. The latter was attended by up to 300 people in Magdeburg, Germany, one of the speakers being the vice president of the so-called European Institute for Climate & Energy (EIKE), Michael Limburg, who has also been an AfD employee, working for the aforementioned Hilse.[31]

Speaking of EIKE points to climate obstructionist think tanks as a class of actors extensively discussed in the previous chapter. However, while we increasingly know about the "denial machine" financed by corporations in the USA, such structures are less developed and researched concerning Europe and the far right.[32] Indeed, such efforts should not be ignored as the networks spreading from the USA, for example via the Heartland Institute or through Atlas Network, enable such actors to flourish.[33] EIKE, to stay with this case, denies the scientific consensus concerning climate change and has significantly engaged with far-right actors. As such, it has received much attention within the far-right media ecosystem. For example, the aforementioned Michael Limburg contributed an article to the far-right monthly *Compact* in January 2020 and its media spokesman Horst-Joachim Lüdecke did even give an interview to the German extreme-right quarterly *Umwelt & Aktiv*.[34] More importantly, EIKE has influenced programmatic developments within the AfD, though its influence on the wider public has been rather limited.[35]

A final example of non-party obstruction concerns social media influencers. Here, the case of the German Naomi Seibt is instructive.

Seibt describes herself as an anarcho-capitalist, but has been called "poster girl of the right" who, when turning 21 in 2021, received public congratulations from, for example, AfD members of the parliament and members of the self-styled "journalistic resistance".[36] Much more, Seibt addresses an international audience through her English language social media activity.[37] Besides the latter, she has spoken in front of the AfD, EIKE and was hired by the Heartland Institute. In fact, she has been described as "a fantastic voice for free markets and for climate realism" by James Taylor from the Heartland Institute and spoke at the so-called Climate Reality Forum running during the COP25 talks in Madrid.[38] Being positioned, and positioning herself, as an "anti-Greta" resulted in much publicity, including in mainstream media. Nevertheless, her impact on public opinion and, especially, the youth is likely to be rather limited, with influencers presenting scientifically established facts being much more widely watched. Indeed, Seibt's influence, like EIKE's, seems to be limited and her YouTube channel was deleted in early 2021 as she spread misinformation regarding COVID-19.[39]

However, the far right is not limited to opposition to evidence concerning anthropogenic climate change. Arguably even more important is their opposition to what academic literature has referred to as process and response scepticism. These "scepticisms" (or "denials") do not require rejection of the scientific evidence but mobilise against climate action by pointing to issues such as the economy and jobs which are arguably more important to people. Let us thus turn to such types of what we call secondary obstruction.

"It's the economy, stupid" (and all this alarmism and hysteria)

Considering what we have outlined so far, there is some truth to the rather widely shared perception that the far right is a literal denialist force. However, research has painted a more complex picture, with work already published during the first half of the 2010s illustrating that stances on human-induced climate change were mixed.[40] This complexity is often overlooked as "extreme" examples are more likely to get the attention of the media, the wider public and researchers – especially if these parties are players in particularly influential countries, such as the AfD in Germany. However, at least as important are those actors who seem to (formally) accept anthropogenic climate change but nevertheless oppose even diluted responses to climate change suggested by scientists, IPCC or centrist and left-wing parties. In other words: such communication strategies, and the wider stories they are connected to, might well accept the scientific evidence – and yet, nevertheless, obstruct necessary climate action. This so-called process and response scepticism, or secondary obstruction in our typology, is what we consider in this section.

Indeed, it has been noted that far-right actors, for example the officially communicated stance of political parties in Nordic countries, have moved away from primary obstruction, though not necessarily embracing policies which will help to counter the climate emergency.[41] While this has long been a feature of obstructionist communication, such secondary forms have become central once again as relevant non-far right communication has moved away from outright denialist arguments, stressing, for example, "costs of climate policies for consumers".[42]

However, before we turn to this major argument revolving around costs, let us consider arguments regarding the process of knowledge generation. This includes accusations of a flawed reviewing processes and the manipulation of evidence, something which rose to particular prominence in the context of the aforementioned "Climategate". Other accusations concern the claim that the left drives the climate agenda as it allegedly fits its ideology and that green industries and climate activists, such as the former U.S. vice president and long-standing climate activist Al Gore, use the debate for their economic benefit. All these arguments can support casting doubt on the very existence of anthropogenic climate change or be used to "simply" call for "a realistic approach" – and they are, thus, known from neoliberal/conservative attempts to delegitimising knowledge about climate change.

Empirical research has furthermore pointed to two related, secondary-obstructionist arguments as standing out. The first one revolves around narratives which thematise the supposed destruction of the nation state by supranational forces, that a New World Order, a world government is around the corner. This claim is arguably particularly relevant within the far-right ecosystem as it is directly connected to its core ideological beliefs. In other words, this claim concerns warnings that processes implemented to address climate change make and are directed towards making national sovereignty redundant – something nationalist actors find, unsurprisingly, difficult to swallow.[43] The fact that such obstruction of processes can often not sharply be demarcated from obstruction of policy responses is visible in a similar, even more banal example taken from another European Parliament contribution, this time by a member of the Freedom Party of Austria who warns that "new institutions will be created which further prune national competences".[44] Relatedly, and before turning to the second major secondary-obstructionist argument, let us point to what has been conceptualised in terms of Ego-Ecology, a concept covering a wider spectrum than only the far right. "Ego-Ecology" is about "egoistic, immediate local or domestic (environmental) demands and interests" as visible, for example, in the case of the French National Rally.[45] Here, environmental/climate protection are not rejected per se but become framed for "selfish" (nationalistic) purposes.

The second major argument concerns the allegedly irrational nature of debates over climate change, of public deliberation being distorted by alarmism, hysteria and the so-called climate religion. This is an argument at times also present in mainstream media, one which indeed has a long tally as it dates back to the 1970s, as discussed in Chapter 2.[46] However, while no systematic comparison between mainstream media and accusations of irrationalism by far-right sources has been published (to our knowledge), it is plausible to assume that the far right puts such arguments forward in a much more antagonising way. This is visible, for example, when it comes to young, female climate activists, such as Greta Thunberg. Here, the accusation of being irrational resonates with longstanding misogynistic perceptions of females as unreasonable and hysteric.[47] Research on far-right media (but also party communication) in Germany and Sweden has carved out precisely such dynamics as Thunberg is ridiculed and dismissed as someone whose position should not be taken seriously.[48] For example, one of Sweden's well-established, far-right media sites writes that Thunberg is scaring people with her "hysterical behaviour".[49] However, such media have also targeted political opponents more generally, such as "leftists", for example in the still relatively under-researched Polish case.[50]

While it is important to note that accusations of alarmism, hysteria and the use of religion metaphors are in no way solely directed against Thunberg in particular or women in general, it is equally important to note that the misogynist, anti-feminist strand in communicating climate change is not limited to such kinds of othering. Instead, some researchers have argued that the very logic of fossil-fuel-driven industrial modernity is bound up with images of what it means to be a man, thus, hampering a transformation of structures to prevent climate catastrophes. For example, some have spoken of "industrial/breadwinner masculinities", which support fossil-fuelled societies and benefit both industrialists and workers, while others have suggested the concept of "petro masculinity" to understand how contemporary fossil fuels are part of the reproduction of identities and "fossilised masculinities".[51]

These masculinities also unfold in what is probably the most common argument in obstructionist communication, that is, that policy responses to climate change will lead to widespread deindustrialisation, the alleged wrecking of our economy and, relatedly, the suffering of "the little people" (see Figure 4.1). In short, "the common people" will pay the price, not "the cosmopolitan elite" as the proposed policy responses will harm the nation. For example, a recent study shows that the AfD – in no way alone in doing so – attacks climate change mitigation policies as doing harm to *our* national interest/economy.[52] Interestingly, Zoltàn Balczó of the Movement for a Better Hungary (Jobbik) agreed that the EU and its member states should meet the goals of the Paris Agreement but similarly warned that this "would decrease the competitiveness of the European industry which is unacceptable".[53] Although not necessarily the case, industries under threat are thought of as "manly" ones, manufacturing dominated by traditional masculinities, of male workers who "do the hard work". True, some of these jobs, in mining or the car and steel industry, which are often unionised and well-paid, are under threat by the need for a green energy transformation towards sustainability. Therefore, it is crucial that such transitions are not left to "creative destruction" but offer opportunities which speak to both economic and cultural needs of

Figure 4.1 From an article entitled "The big rip off" (Compact 7/2019:23, collage: Iris Fischer, material: shutterstock.com). One example for how the far right (not only in Germany) have positioned powerless, ordinary people vis-à-vis a destructive elite while, simultaneously, gendering the debate.[54]

those affected. They must be just transitions, in terms of both redistribution and recognition.

Clearly, this also applies to alternative energy industries, for example the wind energy industry, which currently are only seldomly celebrated by the far right. In fact, there is probably no energy source more regularly despised than wind energy. Why? One reason might precisely be that relevant jobs, although blue collar, are signified as being rather "clean" and "feminine". This might conflict with traditional, masculine images of industrial jobs. However, solar energy, for example, despite similarly lacking the "masculine appeal" of oil, fire and smoke, does not quite provoke the same outrage. Another reason might be that wind is not bounded, does not belong to those who occupy a territory. As such, it is not longed for in the same way as resources which can be hoarded, as national treasures, including coal, gas and oil. Indeed, this is what some have referred to as "The Mythical Body of the Stock".[55] However, water also flows, but hydropower is open to nationalist imagination, and even wind could be incorporated into narratives by the far right. This is most clearly the case in "seafaring nations" or might be accepted for economic reasons, such as in Denmark. Another argument put forward to explain why alternative energy sources are often viewed sceptically is that for them to be truly effective, the integration of national grids into *transnational* connections is needed.[56] While this is true, grids have been connected (though not enough), but oil and gas pipelines have also meant dependence, in the worst case (for those on the far right): dependence on Muslim countries. Against this background, and as sovereignty and the related idea of autarky are central in nationalist thinking, further research is needed to understand why so many in the far right ultimately object to wind energy in particular and alternative energy sources in general. At times, profane (geopolitical) interests, such as in the case of France with its massive nuclear sector and the expanding nuclear sector in Central and Eastern Europe, might play a role, and from time to time even environmental reasons are raised in opposition to alternative energy sources, from soil sealing to the death of birds.[57] In some cases, such environmental claims might be sincere (though probably not in the case of the Finns Party's talk about exploding bats!).[58] However, two reasons are most plausible – and banal: first, these alternatives are thoroughly associated with the political opponent and, thus, are strongly emotionalised.[59] Second, and intertwined with the previous reason, is an ideological one connected to aesthetic concerns and the fact that the far right perceives itself as being rooted in the homeland, as being in a deep (cultural) connection with it. Hence, changes to this landscape must irritate. Of course, not only far-right members of the public have aesthetic concerns when it comes to wind turbines, but the far right in particular might feel the pain as the nation's landscape changes.

A final note on the far right and economy/jobs: although a populist communication strategy revolving around the juxtaposition of "pure people" versus an "evil, corrupt" elite is visible across far-right climate communication, it is maybe with regard to this dimension that it arises most extensively and explicitly. For the far right, it is according to the doings of such an elite that economic competitiveness is lost and hardship for individual members of the nation is caused (unsurprisingly, this elite also threatens national sovereignty and spreads irrationalism). The prevalence of such patterns has led some to look closer at populism as such and its influence on climate change obstruction, arguing that populist, anti-elite stances explain climate obstruction. That is, populism in general is bad for the climate, irrespective of its host ideology, be it left or right.[60] However, while populist elements are certainly mobilised in far-right climate communication, others have shown that climate obstruction is most strongly predicted by exclusionary and anti-egalitarian preferences, nationalist attitudes and xenophobia.[61] Accordingly, a direct comparison of manifestos of right-wing and left-wing populist parties has illustrated the significance of political ideology in relation to climate policy discourse and action.[62]

Moving to another, often-found argument, one connected to concerns over the economy and jobs, there are those lamenting about *us* doing too much too quickly. Consequently, *our* economic power suffers – and *we* need to wait for a global agreement which also binds, especially, the USA, China and India, and forces them to contribute their fair share. However, such arguments concerning *us* doing too much, *us* being held back by climate policies, are neither specific to "the West" nor to the far right. Indeed, we have already seen in Chapters 2 and 3 that Social Democrats in Sweden (with regard to electricity-intense industries) and Norway (with regard to oil extraction) made such arguments too. Today, such arguments can furthermore be found in, for example, China and the Philippines.[63] A related claim concerns the need for pragmatic policy responses – again, this is not specifically far right, but a claim which enables the far right too to articulate its standpoint as a rational standard. This can be thought of as an extension of the above argument, which criticises others for being irrational. Here, however, the focus is on *us* as a nation being "realistic" and rational, on *us* as being able to implement reasonable policy responses.[64]

Looking back, we have seen that although the organisational network of far-right obstruction appears to be far less developed than in
the case of corporations, its partly strategic, partly ideational articula-
tion of climate change situates this obstruction within a far-right (and,
regularly, populist) meaning system. As such, and cutting across primary and secondary obstruction, it is not surprising that research has
also begun to point to race and whiteness within far-right storytelling
about the changing climate. Indeed, the natural environment has long
served as a boundary mechanism, insisting on "our" territory staying
"pure" so as to guarantee the survival of the ethnicised/racialised community – and there is no doubt that the far right might do the same
when it comes to climate change.[65] Indeed, some have argued that at
least in "the West", fossil fuels have been inserted into whiteness so
that, as "whites" are viewed as genuine creators of wealth, climate
change also raises questions regarding privilege.[66] However, the most
straightforward racial (or identitarian) concern at the intersection of
far-right thinking and climate change – nativist arguments to do with
the so-called climate migration – are not yet at the forefront of far-
right imaginations. Indeed, while they are present, work on both the
European Parliament and far-right Austrian media indicates that this
argument remains less than secondary.[67] However, it might become
more central, and not just in communication by the far right.

The far right and climate change: voting behaviour and government activity

So far, we have considered action through talking (and writing), but
the far right has also intervened through other modes, including voting at the national and supranational level as well as, in some cases,
participation in governments. Here, more research is needed – though
effects of the Trump administration in the USA and Bolsonaro in
Brazil are well documented. As we have indicated at the beginning of
this chapter, the former withdrew from the Paris Agreement and sabotaged environmental and climate protection through executive
appointments and cutting relevant budgets, while Bolsonaro did not
withdraw but similarly sabotaged protection through institutional
changes, appointments and financial cuts.[68]

Moving back to the Global North, more specifically, to Europe,
work on the European Parliament rather clearly indicates voting patterns which are in line with the above communication strategies. That
is, although some far-right actors do support climate-related legislation, such legislation is opposed more often than not.[69] For example,

while both the Danish People's Party and the Finns Party voted for climate proposals in a majority of investigated cases during the European Parliament's eight legislative term, the AfD and the Dutch Party for Freedom voted against each and every one of them.[70] But while these two actors have shown a strong dose of primary obstruction, parties like the French National Rally and the Italian Lega, which have also predominantly opposed these proposals, commonly accept the scientific evidence for anthropogenic climate change.[71] At the national level, the ruling Hungarian Fidesz has increasingly paid attention to climate protection and supports the Paris Agreement; however, the Polish Law and Justice party, in government since 2015 and having turned far-right in the course of the so-called refugee crisis of the mid-2010s, has persuaded climate policies of "environmental irresponsibility", while an analysis of legislative motions and party documents by the Freedom Party of Austrian claims to illustrate their aim to combat anthropogenic climate change.[72] Further analysis of populists in government in EU member states reports increasing greenhouse gas emissions, especially in the case of "right-wing populist parties".[73] Others have indicated mixed results, suggesting that government participation of populist radical-right parties does not necessarily lead to massive deterioration of climate policies.[74] This might be because such parties, ultimately, do not prioritise this policy area or that their partners in government insist on some level of seriousness when it comes to this policy field. Finally, and based on a study of OECD countries, Ben Lockwood and Matthew Lockwood have illustrated the negative effect of what they refer to as right-wing populist parties in government on climate policy.[75] However, they also show that this negative effect is significantly moderated by a proportional representation electoral system (that is, not one which hands all power to the victor) and membership in the EU. Furthermore, the negative effect of right-wing populist parties in government on climate policy does not overall extend to renewable policy.

Summary

This chapter aimed to further deepen our understanding of climate obstruction by moving from organised, corporate action to the various modes through which far-right actors, both political parties and non-party actors, have aided climate obstruction. We started with a brief introduction to the far right and why considering this spectrum of actors is useful in the first place. We then turned to arguments and claims made by these actors. In contrast to previous chapters, research covers the period since the mid-/late-2000s and, as such, the primary obstruction we observe during early years – often connected to the so-called Climategate – will have to be contextualised through further, longitudinal studies. While primary obstruction is still visible amongst some of today's far-right actors, maybe most significantly the AfD, secondary obstruction is the norm and relates to, first and foremost, concerns about the plight of the national economy and "the little guy".

What we have not yet covered is what motivates (far-right) actors at the individual level and, more broadly, how to understand what we have called tertiary obstruction. To solve this puzzle, we now turn to Chapter 5 to understand the wider, psychological underpinnings of climate obstruction.

Notes

1 However, and as not only observable in the USA, actors further to the right, such as the so-called alt-right, did not necessarily mirror such climate-obstructing stances. See, for example, Taylor, B. (2019) "Alt-right ecology: ecofascism and far-right environmentalism in the United States" in Forchtner, B. (ed) *The Far Right and the Environment: Politics, Discourse and Communication.* Oxon: Routledge, pp.275–292; Rueda, D. (2020) "Neoecofascism: the example of the United States", *Journal for the Study of Radicalism*, 14(2), pp.95–126. https://doi.org/10.14321/jstudradi.14. 2.0095. Both point out that some in the alt-right have explicitly opposed climate denialism in demarcating themselves from others within the U.S. right. In Europe, Greek party Golden Dawn and the Nordic Resistance Movement (NRM), to name just two, have shown an awareness of anthropogenic climate change too. Such actors blame those who do allegedly not belong to the community for the destruction of the environment. On Golden Dawn, see Forchtner, B. and Lubarda, B. (2022) "Scepticisms and beyond? A comprehensive portrait of climate change communication by the far right in the European Parliament", *Environmental Politics.* https:// doi.org/10.1080/09644016.2022.2048556; on the NRM, see Szenes, E. (2021) "Neo-Nazi environmentalism: the linguistic construction of ecofascism in a Nordic Resistance Movement manifesto", *Journal for Deradicalization*, 27, pp.146–192, and Darwish, M. (2021) "Nature, masculinities, care and the far-right" in Hultman, M. and Pulé, P. (eds) *Men, Masculinities, and Earth Contending with the (m)Anthropocene.* Cham: Palgrave Macmillan, pp.183–206. https://doi.org/10.1007/978-3-030-54486-7_8. On ecofascism and climate change in general, see Moore, S. and Roberts, A. (2022) *The Rise of Ecofascism: Climate Change and the Far Right.* Cambridge: Polity.

2 On Bolsonaro's agenda, see Deutsch, S. (2021) "Populist authoritarian neoliberalism in Brazil: making sense of Bolsonaro's anti-environment agenda", *Journal of Political Ecology*, 28(1), pp.823–844. https://doi. org/10.2458/jpe.2994; Hochstetler, K. (2021) "Climate institutions in

Brazil: three decades of building and dismantling climate capacity", *Environmental Politics*, 30(1), pp.49–70. https://doi.org/10.1080/09644016. 2021.1957614; CSSN (2021) Position Paper 3, "Dismantling the Environmental State: Actors, Strategies and Discourses Behind the Bolsonaro Attack on the National Environmental Regulation"; Menezes, R.G. and Barbosa Jr., R. (2021) "Environmental governance under Bolsonaro: dismantling institutions, curtailing participation, delegitimising opposition", *Zeitschrift für Vergleichende Politikwissenschaft*, 15, pp.229–247. https://doi.org/10.1007/s12286-021-00491-8.

3 For a paradigmatic manifestation of such *ethnopluralism*, see de Benoist, A. and Champetier, C. (1999) "The French New Right in the year 2000", *Telos*, pp.117–144. For a discussion of this manifesto and the French far right from which it emerged and *ethnopluralism*, see Bar-On, T. (2013) *Rethinking the French New Right: Alternatives to Modernity*. Oxon: Routledge.

4 For more on the fourth wave and the idea of far-right waves in general, see Mudde, C. (2019) *The Far Right Today*. Cambridge: Polity.

5 While the classification of individual actors is sometimes controversial, there is, overall, widespread agreement with lists of far-right parties largely identifying the same ones. Besides various monographs and edited volumes, see also Rooduijn, M. *et al.* (2019) "The PopuList: An overview of populist, far right, far left and Eurosceptic parties in Europe". Available at: www.popu-list.org [Last accessed May 3, 2022]; Bakker, R. *et al.* (2020) *2019 Chapel Hill Candidate Survey*, Version 2019.1, Chapel Hill, NC: University of North Carolina. Available at: chesdata.eu [Last accessed May 3, 2022].

6 See Forchtner, B. and Kølvraa, C. (2015) "The nature of nationalism: populist radical-right parties on countryside and climate", *Nature & Culture*, 10(2), pp.199–224. https://doi.org/10.3167/nc.2015.100204 and Lockwood, M. (2018) "Right-wing populism and the climate change agenda", *Environmental Politics*, 27(4), pp.712–732. https://doi.org/10.1080/096440 16.2018.1458411. Of course, this does not imply that arguments concerning economic issues and jobs, might not play a – potentially even *the* – dominant role in far-right communication about climate change as such arguments might emerge within the aforementioned ideological parameters (and speak to interconnected ideological elements, such as populism). Although not singling out the far right, Lübke, C. (2021) "Socioeconomic roots of climate change denial and uncertainty among the European population", *European Sociological Review*, 38(1), pp.153–168. https://doi. org/10.1093/esr/jcab035 points to the relevance of socioeconomic factors. For another study stressing how economic factors affect climate change concerns, see Duijndam, S. and Beukering, P. (2021) "Understanding public concern about climate change in Europe, 2008–2017: the influence of economic factors and right-wing populism", *Climate Policy*, 21(3), pp.353–367. https://doi.org/10.1080/14693062.2020.1831431.

7 Forchtner, B. and Olsen, J. (2023) "Double vision: local *Naturschutz* and global climate change through the German far-right lens" in Forchtner, B. (ed) *Visualising Far-Right Environments. Communication and the Politics of Nature*. Manchester: Manchester University Press.

8 As mentioned in endnote 1 in this chapter, some far-right actors have called for climate action. This can be rooted in calls to protect to "the

local", as when Hungarian Prime Minister Viktor Orbán states: "If we truly fear for our land, our natural environment and our climate, then it is time to act and not simply talk". See Hoerber, T., Kurze, K. and Kuenzer, J. (2021) "Towards ego-ecology? Populist environmental agendas and the sustainability transition in Europe", *The International Spectator*, 56(3), pp.41–55. https://doi.org/10.1080/03932729.2021.1956718.

9 Forchtner, B. (2019) "Climate change and the far right", *WIREs Climate Change*, 10(5), pp.1–11. https://doi.org/10.1002/wcc.604.

10 It is worth noting that the latter group might well accept anthropogenic climate change – though this does not mean that they necessarily support steps towards mitigation proposed by centre-right/left or left-wing/Green parties.

11 On political opportunity structures in general and the far right in particular, see, for example, Spies, D. and Franzmann, S. (2011) "A two-dimensional approach to the political opportunity structure of extreme-right parties in western Europe", *West European Politics*, 34(5), pp.1044–1069. https://doi.org/10.1080/01402382.2011.591096.

12 On the process of framing a message, see Entman's classic definition: "to frame is to select some aspects of a perceived reality and make them more salient in a communicating text, in such a way as to promote a particular problem definition, causal interpretation, moral evaluation, and/or treatment recommendation". Entman, R. (1993) "Framing: towards clarification of a fractured paradigm", *Journal of Communication*, 43(4), pp.51–58. https://doi.org/10.1111/j.1460-2466.1993.tb01304.x.

13 For a useful collection of articles on the BNP, see Copsey, N. and Macklin, G. (2011) *British National Party: Contemporary Perspectives*. Oxon: Routledge.

14 For a discussion of the BNP's climate change communication, see Forchtner, B. and Kølvraa, C. (2015) "The nature of nationalism".

15 For more on this speech and the far right, including the BNP, and climate change in the European Parliament, see Forchtner, B. and Lubarda, B. (2022) "Scepticisms and beyond?" and Schaller, S. and Carius, A. (2019) *Convenient Truths. Mapping Climate Agendas of Right-Wing Populist Parties in Europe*. Berlin: Adelphi.

16 Griffin, N. (2021) Twitter, October 30. Available at: https://twitter.com/NickGriffinBU/status/1454497012720754696?s=20 [Last accessed November 11, 2021].

17 Lewandowsky, S. (2014) "Conspiratory fascination versus public interest: the case of 'Climategate'", *Environmental Research Letters*, 9, 111004. https://doi.org/10.1088/1748-9326/9/11/111004.

18 Dansk Folkeparti (2009) "Invitation to the alternative climate conference", November 13, 2009. Available at: https://danskfolkeparti.dk/invitation-til-alternativ-klimakonference/ [Last accessed September 24, 2021].

19 Forchtner, B. (2019) "Articulations of climate change by the Austrian far right: a discourse-historical perspective on what is 'allegedly manmade'" in Wodak, R. and Bevelander, P. (eds) *"Europe at the Cross-road": Confronting Populist, Nationalist and Global Challenges*. Lund: Nordic Academic Press, pp.159–179.

20 Uscinski, J., Douglas, K. and Lewandowsky, S. (2017) "Climate change conspiracy theories" in Nisbet, M.C. *et al.* (eds) *The Oxford Encyclopedia of Climate Change Communication*. Oxford University Press.

88   *The far right and climate obstruction*

21 van Prooijen, J.-W., Krouwel, A. and Pollet, T. (2015) "Political extremism predicts belief in conspiracy theories", *Social Psychological and Personality Science*, 6(5), pp.570–578. https://doi.org/10.1177/1948550614567356.
22 Ibid., p.572.
23 Fraune, C. and Knodt, M. (2018) "Sustainable energy transformations in an age of populism, post-truth politics, and local resistance", *Energy Research & Social Science*, 43, pp.1–7. https://doi.org/10.1016/j.erss.2018.05.029.
24 AfD (2016) "Program für Deutschland". Available at: https://www.afd.de/wp-content/uploads/sites/111/2018/01/Programm_AfD_Online-PDF_150616.pdf [Last accessed August 30, 2021].
25 AfD (2021) "Programm der Alternative für Deutschland für die Wahl zum 20. Deutschen Bundestag". Available at: https://www.afd.de/wp-content/uploads/sites/111/2021/06/20210611_AfD_Programm_2021.pdf [Last accessed August 30, 2021].
26 AfD (2019) "Alternatives Klimasymposium mit Lord Monckton, Thomas Wysmüller, Prof. Svensmark & Prof. Patzelt", *Youtube*, 15 May. Available at: https://www.youtube.com/watch?v=47ygK7KlzQQ [Last accessed August 12, 2021].
27 Lynch, P., Whitaker, R. and Loomes, G. (2012) "The UK independence party: understanding a niche party's strategy, candidates and supporters", *Parliamentary Affairs*, 65, pp.733–757. https://doi.org/10.1093/pa/gsr042; Turner-Graham, E. (2019) "'Protecting our green and pleasant land': UKIP, the BNP and a history of green ideology on Britain's far right" in Forchtner, B. (ed) *The Far Right and the Environment: Politics, Discourse and Communication*. Abingdon: Routledge. pp.56–71; Bale, T. (2018) "Who leads and who follows? The symbiotic relationship between UKIP and the conservatives – and populism and Euroscepticism", *Politics*, 38(3), pp.263–277. https://doi.org/10.1177/0263395718754718; Forchtner, B. and Lubarda, B. (2022) "Scepticisms and beyond?"; Reed, M. (2016) "'This loopy idea' an analysis of UKIP's social media discourse in relation to rurality and climate change", *Space and Polity*, 20(2), pp.226–241. https://doi.org/10.1080/13562576.2016.1192332.
28 For example, Goodwin, M. (2015) "The Great Recession and the rise of populist Euroscepticism in the United Kingdom" in Kriesi, H. and Pappas, T. (eds) *European Populism in the Shadow of the Great Recession*.
29 Quoted in Forchtner, B. and Lubarda, B. (2022) "Scepticisms and beyond?", p.15.
30 For these and many more examples, see Malm, A. and the Zetkin Collective (2021) *White Skin, Black Fuel: On the Dangers of Fossil Fascism*. London: Verso.
31 See FARN (2019) "Wenn extreme Rechte über Klima reden: Compact-Konferenz 'Gegen den Klimawahn'". Available at: https://www.nf-farn.de/extreme-rechte-klima-reden-compact-konferenz-gegen-klimawahn [Last accessed December 5, 2021]; Compact (2019) "Konferenz gegen den Klimawahn im November 2019". Available at: https://www.compact-online.de/oeko-diktatur-konferenz-2019/ [Last accessed December 5, 2021] and RND (2019) "Ärger im Bundestag: AfD lädt Leugner des Klimawandels als Experten ein". Available at: https://www.rnd.de/politik/arger-im-bundestag-afd-ladt-klimaleugner-als-experten-ein-RNA6EWD6NRADVDEHEUXWJY4TMY.html [Last accessed December 5, 2021]

32 Almiron, N., Boykoff, M., Narberhaus, M. and Heras, F. (2020) "Dominant counter-frames in influential climate contrarian European think tanks", *Climatic Change*, 162(4), pp.2003–2020. https://doi.org/10.1007/s10584-020-02820-4; Busch, T. and Judick, L. (2021) "Climate change – that is not real! A comparative analysis of climate-sceptic think tanks in the USA and Germany", *Climatic Change*, 164(18). https://doi.org/10.1007/s10584-021-02962-z.

33 Plehwe, D. (2022) "Reluctant transformers or reconsidering opposition to climate change mitigation? German think tanks between environmentalism and neoliberalism", *Globalizations*, pp.10–11. https://doi.org/10.1080/14747731.2022.2038358.

34 Compact (2020) "Die CO2-Lüge", *Compact*, pp.44–46; Umwelt & Aktiv (2019) "Im Gespräch mit Prof. Dr. Horst-Joachim Lüdecke vom Europäischen Institut für Klima und Energie (EIKE)", *Umwelt & Aktiv*, 2019(3/4), pp.22–25.

35 Moreno, J.A., Kinn, M. and Narberhaus, M. (2022) "A stronghold of climate change denialism in Germany: case study of the output and press representation of the think tank EIKE", *International Journal of Communication*, 16, pp.267–288. See also Plehwe, D. and Neujeffski, M. (2020) "EIKE: global-warming denial repertoires in action", *Think Tank Network Research Initiative*, March 19. Available at: http://thinktanknetworkresearch.net/blog_ttni_en/denial-repertoires-the-roles-of-germanys-man-made-global-warming-denial-think-tank-eike/ [Last accessed January 5, 2022].

36 Flesch, O. (2021) "Unser Wunderkind ist nun eine junge Dame", *1984 Magazin*, 18 August. Available at: https://19vierundachtzig.com/2021/08/18/nur-das-beste-zum-einundzwan [Last accessed August 18, 2021]. See Buse, U. (2020) "Wie die Einser-Schülerin Naomi zum Postergirl der Rechten wurde", *Der Spiegel*, September 11. Available at: https://www.spiegel.de/politik/deutschland/die-anti-greta-a-00000000-0002-0001-0000-000172993208 [Last accessed September 4, 2021].

37 For more on Seibt's activities, see Buse, U. (2020) "Wie die Einser-Schülerin Naomi zum Postergirl der Rechten wurde".

38 See Butler, D. and Eilperin, J. (2020) "The anti-Greta: a conservative think tank takes on the global phenomenon", *The Washington Post*, 23 February. Available at: https://www.washingtonpost.com/climate-environment/2020/02/23/meet-anti-greta-young-youtuber-campaigning-against-climate-alarmism/?hpid=hp_hp-top-table-low_antigreta-815pm:homepage/story-ans [Last accessed December 5, 2021]; Argüeso, O. and von Daniels, J. (2020), *Correctiv*, February 4. Available at: https://correctiv.org/top-stories/2020/02/04/die-heartland-lobby-2/ [Last accessed December 5, 2021].

39 See Seibt (2021) Twitter, April 30. Available at: https://twitter.com/SeibtNaomi/status/1388130624775544833?s=20 [Last accessed March 29, 2022]; TeamYouTube (2021) Twitter, May 5. Available at: https://twitter.com/TeamYouTube/status/1389853559676604420?s=20 [Last accessed March 29, 2022]. Seibt returned with a new channel shortly before the general election in Germany at the end of September 2021 – with a video entitled "CLIMATE ACTIVISTS ABUSED FOR THE ELECTION – NAOMI SEIBT" [This video is no longer available because the YouTube account associated with this video has been closed].

40 Gemenis, K., Katsanidou, A. and Vasilopoulou, S. (2012) "The politics of anti-environmentalism: positional issue framing by the European radical right", *Paper prepared for the MPSA Annual Conference*, 12–15 April 2012, Chicago, IL; Voss, K. (2014) *Nature and Nation in Harmony: The Ecological Component of Far Right Ideology* (Unpublished doctoral thesis), European University Institute.

41 Kølvraa, C. (2019) "Wolves in sheep's clothing? The Danish far right and 'wild nature'" in Forchtner, B. (ed) *The Far Right and the Environment*. Abingdon: Routledge, pp.107–120; Vihma, A., Reischl, G. and Andersen, A. (2021) "A climate backlash: comparing populist parties' climate policies in Denmark, Finland, and Sweden", *The Journal of Environment & Development*, 30(3), pp.219–239. https://doi.org/10.1177/10704965211027748. For another study on opposition to mitigation measures which includes analysis of far-right parties, see Hess, D. and Renner, M. (2019) "Conservative political parties and energy transitions in Europe: opposition to climate mitigation policies", *Renewable and Sustainable Energy Reviews*, 104, pp.419–428. https://doi.org/10.1016/j.rser.2019.01.019.

42 Cann, H. and Raymond, L. (2018) "Does climate denialism still matter? The prevalence of alternative frames in opposition to climate policy", *Environmental Politics*, 27(3), p.449. https://doi.org/10.1080/09644016.201 8.1439353; Mann, M. (2021) *The New Climate War: The Fight to Take Back Our Planet*. London: Scribe.

43 For example, Mylene Troszcynski, who represents the French National Rally in the European Parliament, has objected to regulation concerning CO_2 emissions as it would empower the European Commission at the expense of member states' sovereignty. Troszcynski, M. (2018) European Parliament designation: (A8-0010/2018) FR 12-06-2018.

44 Obermayer, F. (2015) European Parliament designation: P8_CRE-REV (2015)10-14(1-501-0156).

45 Hoerber, T., Kurze, K. and Kuenzer, J. (2021) "Towards ego-ecology?".

46 Atanasova, D. and Koteyko, N. (2017) "Metaphors in Guardian online and mail online opinion-page content on climate change: war, religion, and politics", *Environmental Communication*, 11(4), pp.452–469. https://doi.org/10.1080/17524032.2015.1024705.

47 Scull, A. (2011) *Hysteria. The Disturbing History*. Oxford University Press. On the link between sexism and climate denial and policy opposition beyond the far right, see, for example, Benegal, S. and Holman, M.R. (2021) "Understanding the importance of sexism in shaping climate denial and policy opposition", *Climatic Change*, 167(48). https://doi.org/10.7910/DVN/EQ8PVR.

48 On far-right non-party media in Sweden and Germany, see Vowles, K. and Hultman, M. (2021) "Scare-quoting climate. The rapid rise of climate denial in the Swedish far-right media ecosystem", *Nordic Journal of Media Studies*, 3(1), pp.79–95. https://doi.org/10.2478/njms-2021-0005; Vowles, K. and Hultman, M. (2022) "Dead White men vs. Greta Thunberg: nationalism, misogyny, and climate change denial in Swedish far-right digital media", *Australian Feminist Studies*. https://doi.org/10.1080/08164649 .2022.2062669; Forchtner, B. (2023) "Thunberg, not iceberg: visual melodrama in far-right climate-change communication in Germany" in Allen, I., Ekberg, K., Holgersen, S. and Malm, A. (eds) *Political Ecologies of the*

Far Right. Manchester: Manchester University Press. On similar anti-Thunberg arguments in party-communication, see, for example, Forchtner, B. and Özvatan, Ö. (2022) "De/legitimising EUrope through the performance of crises: the far-right Alternative for Germany on 'climate hysteria' and 'corona-hysteria'", *Journal of Language and Politics*, 21(2), pp.208–232. https://doi.org/10.1075/jlp.21064.for.
49 Vowles, K. and Hultman, M. (2022) "Dead White men vs. Greta Thunberg".
50 Zuk, Z. and Szulecki, K. (2020) "Unpacking the right-populist threat to climate action: Poland's pro-governmental media on energy transition and climate change", *Energy Research & Social Science*, 66, 101485. https://doi.org/10.1016/j.erss.2020.101485. Similarly, see Bennett, S. and Kwiatkowski, C. (2019) "The environment as an emerging discourse in Polish far-right politics" in Forchtner, B. (ed) *The Far Right and the Environment*. pp.216–236. For a fascinating, ethnographic approach to the Polish far right and its communication about environmental issues, including climate change (obstructionism), see Lubarda, B. (2021) "Far-right ecologism: environmental politics and the far right in Hungary and Poland" (Doctoral thesis), Central European University.
51 Hultman, M., Björk, A. and Viinikka, T. (2019) "Far-right and climate change denial" in Forchtner, B. (ed) *The Far Right and the Environment*. pp.121–135; Daggett, C. (2018) "Petro-masculinity: fossil fuels and authoritarian desire", *Millennium*, 47(1), pp.25–44. https://doi.org/10.1177/0305829818775817; Malm, A. and the Zetkin Collective (2021) *White Skin, Black Fuel*. For an excellent analysis of coal miners' "industrial breadwinning petro-masculinities" in Upper Silesia, Poland and the allures of the far right under such conditions of economic uncertainty and gender anxiety, see Allen, I.K. (2022) "Heated attachments to coal: everyday industrial breadwinning petro-masculinity and domestic heating in the Silesian home" in Iwińska, K. and Bukowska, X. (eds) *Gender and Energy Transition. Case Studies from the Upper Silesia Coal-Mining Region*. Cham: Springer, pp.189–222. Similarly insightful is Kojola, E. (2019) "Bringing back the mines and a way of life: populism and the politics of extraction", *Annals of the American Association of Geographers*, 109, pp.371–381. https://doi.org/10.1080/24694452.2018.1 506695 on the rural Iron Range mining region in northern Minnesota, USA, which illuminates how place-based identities and moral economies tied to mining are central to right-wing populism. On climate denial, gender and ontological insecurity, see also Agiusa, C., Rosamond, A.B. and Kinnvall, C. (2020) "Populism, ontological insecurity and gendered nationalism: masculinity, climate denial and Covid-19", *Politics, Religion & Ideology*, 21(4), pp.432–450. https://doi.org/10.1080/2156768 9.2020.1851871; On the misogynist construction of "irrational women" by the Finish far right, see Pettersson, K., Martikainen, J., Hakoköngäs, E. and Sakki, I. (2022) "Female politicians as climate fools: intertextual and multimodal constructions of misogyny disguised as humor in political communication", *Political Psychology*. https://doi.org/10.1111/pops.12814.
52 Küppers, A. (2022) "'Climate-Soviets,' 'alarmism,' and 'eco-dictatorship': the framing of climate change scepticism by the populist radical right

alternative for Germany", *German Politics*. https://doi.org/10.1080/09644 008.2022.2056596.

53 Voigt, U. (2017) European Parliament designation: (B8-0534/2017). Available at: https://www.europarl.europa.eu/meps/de/124832/UDO_VOIGT/ all-activities/written-explanations/8 [Last accessed March 30, 2022].

54 Image taken from Meissner, K. (2019) "Die große Abzocke", *Compact*, 7, pp.23–25. For a detailed analysis of this image, see Forchtner, B. (2023) "Thunberg, not iceberg".

55 Malm, A. and the Zetkin Collective (2021) *White Skin Black Fuel*, pp.274–279.

56 Ibid.

57 See Mann, M. (2021) *The New Climate War*, p.128. With regards to nuclear energy, the Swedish case furthermore illustrates significant nostalgic attachment to this technology.

58 Hatakka, N. and Välimäki, M. (2019) "The allure of exploding bats: the Finns Party's populist environmental communication and the media" in Forchtner, B. (ed) *The Far Right and the Environment*. pp.136–150.

59 For example, it has been shown that negative attitudes towards the energy transition and wind turbines in Germany are associated with support for the AfD, and that, at the municipality level, the party benefits (in terms of electoral success) from the construction of wind turbines. See Otteni, C. and Weisskircher, M. (2021) "Global warming and polarization. Wind turbines and the electoral success of the greens and the populist radical right", *European Journal of Political Research*. https://doi.org/10.1111/ 1475-6765.12487.

60 Huber, R. (2020) "The role of populist attitudes in explaining climate change skepticism and support for environmental protection", *Environmental Politics*, 29(6), pp.959–982. https://doi.org/10.1080/096440 16.2019.1708186; Böhmelt, T. (2021) "Populism and environmental performance", *Global Environmental Politics*, 21(3), pp.97–123. https://doi. org/10.1162/glep_a_00606; Huber, R.A., Greussing, E. and Eberl, J.M. (2021) "From populism to climate scepticism: the role of institutional trust and attitudes towards science", *Environmental Politics*. https://doi. org/10.1080/09644016.2021.1978200. More generally, Kroll, C. and Zipperer, V. (2020) "Sustainable development and populism", *Ecological Economics*, 176, 106723. https://doi.org/10.1016/j.ecolecon.2020.106723 show that the analysed countries' performance on the 17 Sustainable Development Goals relates to electoral support for populist parties.

61 Jylhä, K. and Hellmer, K. (2020) "Right-wing populism and climate change denial: the roles of exclusionary and anti-egalitarian preferences, conservative ideology, and anti-establishment attitudes", *Analyses of Social Issues and Public Policy*, 20, pp.315–335. https://doi.org/10.1111/ asap.12203; Kulin, J., Johansson Sevä, I. and Dunlap, R. (2021) "Nationalist ideology, right-wing populism, and public views about climate change in Europe", *Environmental Politics*, 30(7), pp.1111–1134. https://doi.org/10.1080/09644016.2021.1898879; Yan, P., Schroeder, R. and Stier, S. (2021) "Is there a link between climate change scepticism and populism? An analysis of web tracking and survey data from Europe and the US", *Information, Communication & Society*, 176, 106723. https://doi. org/10.1080/1369118X.2020.1864005; Krange, O., Kaltenborn, B. and

Hultman, M. (2019) "Cool dudes in Norway: climate change denial among conservative Norwegian men", *Environmental Sociology*, 5(1), pp.1–11. https://doi.org/10.1080/23251042.2018.1488516; Recently, Krange, O., Kaltenborn, B.P. and Hultman, M. (2021) "'Don't confuse me with facts' – how right wing populism affects trust in agencies advocating anthropogenic climate change as a reality", *Humanities and Social Sciences Communications*, 8(55). https://doi.org/10.1057/s41599-021-00930-7 have argued that lack of trust in environmental institutions is associated with evidence scepticism, though immigration scepticism too has a negative effect on trust.

62 Huber, R., Maltby, T., Szulecki, K. and Ćetković, S. (2021) "Is populism a challenge to European energy and climate policy? Empirical evidence across varieties of populism", *Journal of European Public Policy*, 28(7), pp.998–1017. https://doi.org/10.1080/13501763.2021.1918214.

63 Liu, J. (2015) "Low carbon plot: climate change skepticism with Chinese characteristics", *Environmental Sociology*, 1(4), pp.280–292. https://doi.org/10.1080/23251042.2015.1049811; Marquardt, J., Oliveira, C. and Lederer, M. (2022) "Same, same but different? How democratically elected right-wing populists shape climate change policymaking", *Environmental Politics*. https://doi.org/10.1080/09644016.2022.2053423.

64 Such "front runner" rhetoric of a successful nation has dominated the story of a country such as Sweden during the years 2006–2009 with the consequence that per capita emissions was dismissed from the debate. See Anshelm, J. and Hultman, M. (2014) *Discourses of Global Climate Change*. London: Routledge.

65 On nature and the natural environment as a bordering mechanism, see Olsen, J. (1999) *Nature and Nationalism*. New York: St. Martin's Press; Hultgren, J. (2015) *Border Walls Gone Green: Nature and Anti-Immigrant Politics in America*. Minneapolis: University of Minnesota Press; Turner, J. and Bailey, D. (2021) "'Ecobordering': casting immigration control as environmental protection", *Environmental Politics*, 31(1), pp.110–131. https://doi.org/10.1080/09644016.2021.1916197; Forchtner, B. (2019) "Nation, nature, purity: extreme-right biodiversity in Germany", *Patterns of Prejudice*, 53(3), pp.285–301. https://doi.org/10.1080/0031322X.2019.1592303.

66 Malm, A. and the Zetkin Collective (2021) *White Skin, Black Fuel*, pp.315–391.

67 Forchtner, B. (2019) "Articulations of climate change by the Austrian far right" and Forchtner, B. and Lubarda, B. (2022) "Scepticisms and beyond?".

68 Marquardt, J., Oliveira, M.C. and Lederer, M. (2022) "Same, same but different?".

69 Schaller, S. and Carius, A. (2019) *Convenient Truths*. For a more recent study which considers the entire political spectrum in the European Parliament, see Buzogány, A. and Cetkovic, S. (2021) "Fractionalized but ambitious? Voting on energy and climate policy in the European Parliament", *Journal of European Public Policy*, 28(7), pp.1038–1056. https://doi.org/10.1080/13501763.2021.1918220.

70 Schaller, S. and Carius, A. (2019) *Convenient Truths*, p. 27.

71 Forchtner, B. and Lubarda, B. (2022) "Scepticisms and beyond?". For a comparison between the evidence-sceptic AfD and the National Rally, see

Oswald, M., Fromm, M. and Broda, E. (2021) "Strategic clustering in right-wing-populism? 'Green policies' in Germany and France", *Zeitschrift für Vergleichende Politikwissenschaft*, 15, pp.185–205. https://doi.org/10.1007/s12286-021-00485-6.

72 Hoerber, T., Kurze, K. and Kuenzer, J. (2021) "Towards ego-ecology?", p.48; Riedel, R. (2021) "Green conservatism or environmental nativism? Exploring carbon dependency as populist political strategy in Poland", *Zeitschrift für Vergleichende Politikwissenschaft*, 15, pp.207–227. https://doi.org/10.1007/s12286-021-00490-9; Voss, K. (2019) "The ecological component of the ideology and legislative activity of the Freedom Party of Austria" in Forchtner, B. (ed) *The Far Right and the Environment*. pp.153–183.

73 Jahn, J. (2021) "Quick and dirty: how populist parties in government affect greenhouse gas emissions in EU member states", *Journal of European Public Policy*, 28(7), pp.980–997. https://doi.org/10.1080/135017 63.2021.1918215. Similarly, Böhmelt, T. (2021) "Populism and environmental performance" finds that populist leadership is strongly linked to lower environmental performance (as captured by CO_2 emissions), but identifies "only weak evidence for an influence of political ideology". Interestingly, Duijndam and Beukering (2021) find no strong direct relationship between climate sceptic 'right-wing populist parties' and public concern about climate change.

74 Ćetković, S. and Hagemann, C. (2020) "Changing climate for populists? Examining the influence of radical-right political parties on low-carbon energy transitions in Western Europe", *Energy Research & Social Science*, 66, 101571. https://doi.org/10.1016/j.erss.2020.101571; Holgersen, S. (2023) "Delayers and deniers: the Norwegian far-right meeting the centrist oil-ideology" in Allen, I., Ekberg, K., Holgersen, S. and Malm, A. (eds) *Political Ecologies of the Far Right*. Manchester: Manchester University Press.

75 Lockwood, B. and Lockwood, M. (2022) "How do right-wing populist parties influence climate and renewable energy policies? Evidence form OECD countries", *Global Environmental Politics*, 22(3), pp.12–37. https://doi.org/10.1162/glep_a_00659.

5 The public and climate obstruction

Introduction

Previous chapters described how socio-political processes and concerted actions have been playing their part in the formation (and transformation) of the various forms of climate obstruction. We have explained how powerful global and societal actors, such as fossil fuel companies and political parties, have successfully worked against climate initiatives and have substantially delayed mitigation efforts. Consequently, global greenhouse gas emissions have continued to rise despite international climate agreements and the seemingly widespread awareness of the climate crisis.[1] The situation necessitates extensive reforms in society and, thus, neither responsibilities nor solutions are to be found solely in individual action. That is, voluntary behavioural changes alone do not provide sufficient means to address the climate crisis, and individual action is enabled and constrained by various structural and institutional conditions, such as high-carbon infrastructures.[2] However, individuals can collectively contribute to mitigation through their various roles: as ordinary citizens, members of communities, political actors, normative influencers and consumers.[3] Perhaps most importantly, the public can put pressure on politicians and industries that have thus far failed to deliver effective solutions.[4] Indeed, in democratic societies, wide-ranging system changes cannot be implemented without public approval.[5] As writer Mary Annaïse Heglar has pointedly said, "[I]t's true that you can't solve the climate crisis alone, but it's even more true that we can't solve it without you".[6]

After decades of active lobbying and disinformation campaigns, opposition to climate change initiatives is persistent. Many are unwilling to approve the proposed climate policies or to accept the eventual costs of climate mitigation, and some even dismiss the scientific findings that are today considered as facts about climate change.[7] The

DOI: 10.4324/9781003181132-5

public can also unintentionally contribute to delaying mitigation efforts through their collective (in)actions. Disinformation is an important source for the variation in climate-related perceptions, but parts of the public can also be personally motivated to obstruct climate policies and to reproduce the false claims that the denial machine fabricates.[8] Thus, as we argued in Chapter 1, an integrated perspective on climate obstruction is needed. This recognises the value of individual engagement in mitigation efforts but also the embeddedness of individual opinions and actions in a web of socio-economic relationships, information environment and societal structures.

This chapter examines contributions to obstruction among the public, with a mainly social psychological perspective,[9] but drawing from various social sciences. Using our framework, the three forms of obstruction, introduced in Chapter 1, we first shed light on *tertiary obstruction*. Here, we focus on inaction among the worried segments of the public, and particularly discuss attitude-behaviour gaps and the interaction between individual preferences and structural obstacles. Second, we examine *primary obstruction* by describing how basic cognitive biases and motivated reasoning can hamper correct or complete conclusions on climate science. And finally, particularly in relation to *secondary obstruction*, we take a deeper look into social processes, identities and individual difference factors that influence opinions on climate policies and other mitigation efforts. In the concluding part of this chapter, we discuss the unique and common themes identified across the different forms of obstruction.

Attitude-behaviour gap and structural obstacles

Scientific evidence for climate change is not new, but the issue has remained relatively distant for the public in several parts of the world.[10] Yes, climate change has been seen as a serious threat for the future and for some parts of the world in particular, but other issues – that an individual may feel requiring more immediate attention – have often been prioritised in lifestyle choices and voting preferences.[11] There has also been a "spiral of silence" around climate change, meaning that many are concerned about climate change but at the same time say that they do not talk or hear about the issue often.[12]

Several signs indicate that this spiral is breaking and that expressions of mitigation support are growing, at least in some societies. Perhaps the most visible sign of this is the mass demonstrations, particularly the Fridays for Future ones. Moreover, in a 2021 Special Eurobarometer, for the first time in measurement history, climate change was named as the most serious global problem more commonly (18 per cent) than any other issue.[13] The results also showed that the perception of climate change being a serious problem is widespread (93 per cent) and that more than half stated they had personally acted on climate over the past six months (64 per cent). When asked about the current national actions to tackle climate change, 75 per cent believed they are not enough, while only 3 per cent responded that their government is doing too much. Similar trends are observed in the "Global Warming's Six Americas" measurement of the Yale Program on Climate Change Communication, where the alarmed segment of the public has nearly doubled between 2017 and 2021 (from 18 to 33 per cent) – although it should be noted that this is mostly due to the "concerned" part decreasing from 32 per cent to 25 per cent and moving to a more alarmed direction.[14] Nonetheless, also the prevalence of dismissive (9 per cent) and doubtful (10 per cent) show a decreasing trend (from 11 and 12 per cent, respectively). As to global trends, certain conclusions cannot be drawn due to the Western bias in research,[15] but surveys show global concern. For example, in a 2021 survey "The Peoples' Climate Vote" that covered 50 countries with 1.2 million respondents, 64 per cent think that climate change is a global emergency.[16] When asked which policies their country should pursue, out of a list of 18 policies, 97 per cent of respondents supported at least one policy, and the average number of supported policies was 8. In sum, it seems that worry and interest in climate change are becoming more widespread, giving the decision makers a strong mandate to commit to a politics that will dramatically lower emissions. Still, climate obstruction is maintained through individual contributions in various ways, and the level of worry has not resulted in extensive climate action.

One of the puzzles in environmental psychology has been why we are not seeing the extensive behavioural changes and strong demands of political reforms that could be expected based on the expressions of concern and interest. Indeed, there is a historically persistent attitude-behaviour (also called value-action) gap, whereby the environmental knowledge, awareness and concerns are not translated to corresponding pro-environmental behaviour (i.e., behaviour whereby an individual consciously seeks to minimise their negative environmental impact on the natural and built world).[17] The explanations to why individuals do or do not engage in pro-environmental behaviour is complex, including various psychological, economic and structural factors and their interplay, such as sense of personal responsibility, knowledge of the issue, moral and social norms, intention to act, sense of efficacy, and the necessary resources and infrastructure that enable action, to name a few.[18] Some of these factors will be discussed in more detail later in this chapter, but before that we highlight some important aspects in the research and interpretations of the results related to climate action.

One reason for the attitude-behaviour gap in climate action could be related to measurement. Self-reported climate behaviours are in some research only weakly linked to actual behaviour, which in part is due to the difficulties selecting and formulating the core questions.[19] There are various ways to act on climate, and individuals do not typically engage in all of them. It is difficult to capture this variability in questionnaires. It is also socially desirable, easy and cheap to express concern and to support climate mitigation, while the behaviours and policies tend to require lifestyle changes and entail increased costs in terms of economic investments, time consumption or in some situations even social relationships.[20] Basically, people can thus say one thing when asked but then their behaviours do not always follow. To address these limitations, some scholars have proposed that the actual levels of pro-environmental attitudes and concerns should not be estimated by directly asking about them, but by observing what behaviours are displayed. That is, if people only engage in easy and non-costly behaviours, their pro-environmental attitudes and values can be weak, while strong attitudes would likely entail investing more to behaviours, resulting in environmental actions regardless of their difficulty and costs.[21]

Indeed, some important measures of decreasing personal emissions are met with controversy and resistance, which calls into question the actual prevalence of high climate concern. As the Eurobarometer (2021) showed, people are readily engaging in actions such as recycling and cutting down on the consumption of disposable items.[22] These actions are of importance in general environmental protection (e.g., managing the levels of microplastic in ecosystems) but have very limited effect on greenhouse gas emissions. Meanwhile, more important actions, such as decreasing meat consumption, or avoiding flying or driving a car, are less common and have been met with stronger resistance.[23] But how much do these reactions say about lacking high climate concern? While there are correlations between lower concern and such resistance, the correlations should be interpreted with caution.[24] Some of the high-impact actions have only recently been more widely brought up as solutions in public discussions and education systems.[25] These actions are still part of certain infrastructures and admirable consumption patterns and lifestyles, and such actions are difficult for individuals to change regardless of their environmental concern.[26] People may also be concerned about a free-ride effect and thus be more prone to engaging in climate action if they can be sure that others also do their part (e.g., through environmental laws).[27] This response is not surprising since individual actions have a notable impact only if many individuals commit to them: it is possible that, in the cases where climate action requires personal sacrifices, people resist making these changes unless others make them too. It should also be noted that individual consumption has a very limited effect on mitigation efforts and may not be the form of climate action that should be mostly focused on in research and in climate communication. Rather, it could be beneficial to support grassroots engagement, whereby the public can signal both acceptance of – and demand for – effective climate policies and regulations.[28] Such activities are not very common despite the visibility of recent climate demonstrations. For example, a study in Finland found that engagement in civic activism or organisational activities (7 per cent each) are rare and still have a large potential to increase thereby making a big impact.[29]

Nevertheless, collective efforts in changing consumption patterns can also play an important part in climate change mitigation, particularly in wealthy countries where the per capita emissions are high, and among the parts of the public who have high personal emissions. As shown in Figure 5.1, the richest 10 per cent of the global population are responsible for over 50 per cent, and the poorest 50 per cent for only seven per cent, of the global population. This highlights that the responsibility to make behavioural changes is not equally spread across people.

So, what could explain why effective climate action is not more widespread (in terms of high-impact lifestyle choices or grassroot engagement)? Among the most powerful explanations are perhaps the logics of our current petroculture society and how fossil fuels have become intertwined with our comforts and customs.[31] This equation of fossil fuels and the good life has been actively constructed and enforced by, for example, car companies and airlines. This helps to explain why transition away from fossil fuels, and why solutions such as electrification of the car fleet seem more plausible and preferable to many than, for example, the introduction of large public transportation systems. Since the 1960s, many societies have built their transportation infrastructure around the expectation that every family owns a car and that is not easily reversed. Similarly, meat has become integrated into values and culture through social customs and is an inherent part of everyday dishes and festive food, ranging from minced meat and kebab to Christmas ham. Not so long ago, meat was a luxury part of the dish but is today often seen so central part of the meal that vegetarian or vegan foods tend to be classified

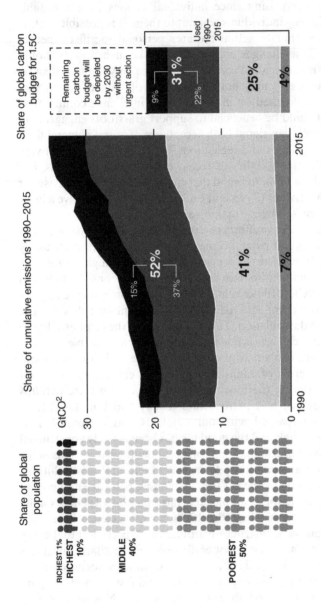

Per capita income threshold ($PPP2011) of richest 1%: $109k; richest 10%: $38k; middle 40%: $6k; and bottom 50%: less than $6k
Global carbon budget from 1990 for 33% risk of exceeding 1.5C: 1,205Gt.

Figure 5.1 Share of cumulative emissions from 1990 to 2015 and use of the global carbon budget for 1.5C linked to consumption by different global income groups.[30]

Source: reproduced from Oxfam, 2020.

as special diets. The meaning of meat goes thus beyond its nutritional value, and there is resistance to plant-based diets.[32] Indeed, vegetarianism is sometimes seen as a threat to culinary traditions and one's country's cultural customs.[33] Meat is also intertwined with societal views on masculinity, which forms an additional obstacle for selecting plant-based options among men.[34]

An interesting case of the attitude-behaviour gap is shown in Kari Marie Norgaard's pioneering study of how a small Norwegian town responded to the issue of climate change. While the country is generally regarded as a forerunner in international environmental governance, it is also a country where the expansive welfare policies are dependent on oil extraction realised through the work of the state-owned company Equinor (previously Statoil). Despite a general pro-environmental attitude with many people having direct relations to farming and an outdoor lifestyle, few considered their life or actions as relevant for climate change.[35] To Norgaard's surprise, this happened despite public engagement in politics in general and despite visible evidence of changes in climate that had a direct impact on the life in the town, including diminishing snowfall hurting the ski-resort. The culture enabled citizens to use a wide variety of tools to deny or fail to act on climate change, for example placing the problem elsewhere.[36] This points to the insight that most of us in the world know to some extent what is going on, but we fail to act either of refusal, denial, apathy or simply because we don't feel we have the resources to make a difference.

Norgaard's analyses illustrate a state where individuals' values, attitudes, beliefs and/or behaviours are inconsistent.[37] Such inconsistency tends to cause cognitive dissonance, a state of mental discomfort that people tend to try to ease somehow.[38] Relief can be achieved by seeking more consistency through altering one of the views or behaviours. In the climate context – where it often is difficult and unpleasant to make concrete changes or influence the situation at a large scale – people may deny climate change or rationalise and justify their actions in other ways.[39]

Importantly, and also in relation to the findings of Nordgaard described above, a sense of efficacy is central: if people do not feel that their behaviour will make a difference (i.e., have a sense of response efficacy), that they have the ability to engage in the required behaviour (self-efficacy), and/or that the collective effort is possible and will be successful (collective efficacy), it can be difficult to find motivation to action.[40] Climate threat can evoke many difficult emotions, and coping with them can be directed to action that aims to avoid the danger, or to alter the appraisal of the threat. If there seem to be no effective ways of acting, people may aim to maintain wellbeing through psychological distancing, avoidance or denial. People can also try to find reasons to be hopeful, and there are concerns that this can lead to decreased acknowledgement of the crisis and to weakened motivation to climate engagement.[41] However, hope is a complex emotion and can be based on various sources. In the climate context, both wellbeing and climate engagement have been connected to so-called constructive hope, which entails an ability to switch between worried thoughts and some form of positive reappraisal, for example trust in different societal actors' willingness and ability to find solutions to the crisis.[42] This differs crucially from forms of hope that are based on wishful thinking and de-emphasising the threat.

Notably, in addition to climate obstruction that is happening among the concerned segments of the public, there still exists outright denial and blatant opposition to climate mitigation. The remaining parts of this chapter focus on explaining reasons for this, starting from the basic cognitive factors and motivated reasoning that influence interpretations of climate knowledges, and continuing to discuss how identities, ideologies and personality dispositions shape preferences and engagement in climate mitigation.

Epistemic and interpretative disagreement on climate change

At the surface, it seems that acknowledgement of the climate crisis is commonly accepted across society – not only in climate science but also in politics, media and business life. Also, the messages from climate scientists are consistently alarming and demand rapid action. What could thus explain why so many still downplay the threat or even doubt the very existence of climate change?

Climate obstruction can take several forms, and what we have called primary obstruction does still exist among the public. For example, in a recent nationally representative study in Sweden – a country where outright denial is relatively rare[43] and which is typically placed among the top countries in climate change mitigation efforts[44] – only 34 per cent of respondents *fully* agreed that climate change is mainly caused by human activities (score 10 on a scale from 1 to 10).[45] Nevertheless, the majority scored between six and ten (81 per cent) rather than between 0 and 5 (19 per cent). Research typically finds that denial (or scepticism) of the trend (existence of climate change) is uncommon, while denial of the attribution (human causes) or the seriousness of climate change is somewhat more widespread – although the prevalence of different forms of denial varies across the cultural contexts.[46]

One could argue that much of the information and perceptions of climate change are objective and should be processed on the basis of scientific evidence. But processing of scientific evidence is not immune to the multiple motivations and cognitive biases that can distort information processing also in general. Our reality is immensely vast, and we are constantly receiving an endless flow of information through our senses, bodily functions and thinking processes that evoke emotions and ideas that further influence us. When making sense of this complex mixture of reality and internal input, it is common to use mental shortcuts.[47] Humans have a limited cognitive capacity and may thus not engage in thorough information processing unless there are important reasons to do so. Mental shortcuts spare time and, in many cases, enable forming good-enough decisions and conclusions. In some other situations, such as it is the case with erroneous conclusions about climate change, the consequences can be serious.

Furthermore, there are some aspects of climate science – or perhaps any science for that matter – vulnerable to questioning and suspicion. Some part of the difficulty comes from the term itself: Climate refers to long-term average weather and temperature patterns in a particular area. These may be observed by humans; after all, we are generally considered to be "naïve researchers", who aim at detecting causal processes and predicting future events.[48] However, while weather patterns are indeed commonly observed and discussed, we are not equipped to intuitively detect changes in the climate because this requires systematic measurements and statistical analyses. Thus, the public depends on climate scientists and experts to provide information and clarify the research results.[49] Moreover, trust in scientist and the scientific process is essential. This gives room for using fake experts – one of the key tactics of science denial as discussed in previous chapters – to confuse public opinion.[50] Each research finding entails some level of scientific uncertainty, which can be incorrectly interpreted as a sign of disagreement or doubt on the conclusions.[51]

The human drive for understanding is actively used to counter current climate science.[52] A sceptical and critical mindset is considered desirable in society, because true (scientific) scepticism treats information as unreliable unless it is supported by reliable evidence.[53] Disinformation on climate change is commonly shaped to be pseudoscientific, which means that it does not meet the scientific credibility standards but nevertheless *appears* to be scientific. At the same time, the countermovements have portrayed climate scientists as corrupt, biased and methodologically incompetent. The uncertainties in climate science have been pointed out, misinterpreted and falsely claimed of being indicators of unreliability and disagreement. Climate concern has been framed as alarmism that is detrimental to mental health and society. In sum, several approaches have been taken to conceptualise climate obstruction as a demonstration of scientific reasoning and healthy scepticism, while agreement and worry are portrayed as emotion-driven and irrational responses.[54] As a result, it can be difficult for laypeople to assess the reliability of the various arguments and information sources, and therefore many may sincerely consider the issue to be unsettled.[55] As we have shown above, this is the precise effect that was sought for by industry actors as exemplified in the internal memo from Exxon stating that victory will be achieved when: "Average citizens 'understand' (recognize) uncertainties in climate science; recognition of uncertainties becomes part of the "conventional wisdom".[56]

However, there is more to the explanation on how climate information is processed. It is true that there is plenty of skilfully crafted disinformation in circulation and this creates confusion. But not everyone believes the same messages. One reason for this is motivated reasoning, which means that the goal of information processing is not necessary to draw the most objective conclusions but is often motivated by the desire to find evidence that confirms some pre-decided assumption.[57] This tendency is related to confirmation bias, which refers to the inclination to trust and believe more such information that confirms one's pre-existing views than information that challenges them.[58] In addition to these more general cognitive biases, the dismissal of climate change can be appealing due to solution aversion.[59] Thus, people may refuse to accept the reality of climate change because they dislike the proposed mitigation policies and required societal reforms. This shows the importance of how climate change is, and has been communicated, and in what way it is constructed as a problem. As shown in Chapters 2 and 3, the portrayal of climate action as solely a cost with detrimental effects on everyday life and threatening the very idea of the good life has been actively promoted and constructed by fossil fuel companies and other actors through the use of specific economic models.[60] This portrayal of for example jobs versus the environment has also aided in gathering support for various astro-turf campaigns and been a key claim by the far right as introduced in Chapter 4.

Identities and ideological resistance to climate change mitigation

The previous chapters described the roles of, for example, vested socio-economic interests and political actors – including the far-right

– in shaping and maintaining climate obstruction. Not surprisingly, then, similar connections have been found among the public in many countries: some of the most consistent and strongest correlates of the various forms of obstruction are political orientation and identities.[61] Right-leaning and conservative segments of the public are more likely to deny climate change, oppose climate policies, and engage in personal and social behaviours that contribute to obstruction. In the Western context, the connection tends to be consistent but varies in strength across countries.[62] For example, it is stronger in the USA than in other countries. However, at least in some non-Western countries, and in former Communist countries in Europe, a similar divide is not found.[63] Moreover, even in cultural contexts where the divide exists, it is more notable among respondents who have strong interest in politics,[64] and who see the environment as a left-right issue.[65] The connection thus depends on the cultural context, for example the strength of political salience of the climate issue, political identity processes and polarisation, and the meaning of left-right identification.

In many countries, conservative political and media have expressed less concern over climate change than liberals.[66] Furthermore – as described in previous chapters – the organised efforts to delay climate action have connections to conservative movements and think tanks[67], which has likely influenced the rhetoric of obstruction and thereby the identification with the arguments. In the situation where a partisan identity has been formed, loyalty for party may override the importance of truth if these two clash.[68] There is evidence for complementary issue alignment in Europe, where many far-right parties have entered the political sphere and aimed to position themselves in relation to other parties.[69] The European far-right promotes a range of sociocultural issues, most importantly opposition to immigration and multiculturalism, and are often placed along the opposite end from green parties. Green parties are thus their natural political rivals, which may in part explain the findings showing that the correlation between anti-environmental attitudes and anti-immigration stances has become stronger over the past decade in some cultural contexts.[70] The group norms and identities that are formed in such processes can limit environmental behaviours among the conservative and (far) right-wing segments of the public and evoke ideological resistance and a sense of ingroup threat in response to environmental messages and proposals.[71]

Another identity process to be considered is gender socialisation. As compared to a traditional upbringing of women, men are socialised to a lesser extent to – or even punished for – being considerate, kind and sensitive, which could in part explain why they tend to express less concern for environment.[72] Male norms and masculine ideals also entail more practices that are now required to change: big cars, meat (vs. being a "soy boy") and gasoline smell. Moreover, in the hierarchical structures of society, nature is placed below humans, and women are placed below men. Indeed, connections have been found between climate obstruction, sexist and anti-feminist attitudes, and views of women being closer to nature.[73] There are thus implicit and explicit environmentally relevant expectations of what being male means, regulated with social punishments and risk of status loss when being perceived as feminine.[74]

However, not all obstructions stem from identity processes and exposure to disinformation. Attitudes are influenced both by "top-down" (from outside, e.g., political communication) and "bottom-up" (from within, e.g., personality dispositions) factors, and these also interact.[75] In line with this, people tend to support parties and politicians with whom they consider sharing personality characteristics and values,[76] and there are also limitations in how much politicians can shape their supporters' views.[77] Thus, it seems possible that, for example, economic interests help explain why conservative politicians and think tanks have adopted anti-environmental agendas, but they may also explain how individuals reason around the issue regardless of the cues that they get from political elite. For example, an individual who strongly opposes governmental regulations would likely oppose environmental laws regardless of being exposed to politicians' statements about the issue.

As mentioned earlier, the politicisation of climate change likely reflects a high degree of solution aversion, whereby certain – or sometimes all – forms of mitigation behaviours and policy suggestions clash with the views of the ideal society and policies.[78] At a broader level, this solution aversion can result in a general dislike of all types of climate discourses, thus explaining climate obstruction in its all (or most) forms. Indeed, measures that capture political issue preferences (e.g., on taxes or immigration) are more consistently connected with environmental views than the more general measures for partisanship and political orientation.[79] However, in this context, it is important to note that people may support some policies while opposing others, and understanding of the more specific forms of opposition is needed to promote political reforms. For example, right-wing ideology is more strongly linked to opposition of policies that are juxtaposed with economic growth,[80] suggesting that when the solutions are connected specifically to some issue that conflicts with the core issues of conservative ideology, polarisation on climate change is particularly strong.

There are also more basic ideological processes that could explain obstruction, beyond the perceived compatibility of climate policies and other political views. Climate change is caused by humans and our contemporary lifestyles. Both climate emergency and mitigation efforts thus challenge our societal structures and traditions, which may jeopardise the sense of continuity and stability. These ideas can be particularly threatening to conservative ideologies, that is, to protection of traditions and conventional values[81] and the societal power structures.[82] Adherence to the system provides sense of stability and security, and helps creating and supporting wellbeing and self-esteem.[83] Thus, it can be difficult to admit the consequences of the current practices on our planet, and the need to make a change.[84] Indeed, different forms of obstruction have been associated with conventionalism and desire to protect status quo.[85] Importantly, yet another, more specific, form of conservative ideology has been linked to obstruction: Social Dominance Orientation.[86] The latter refers to a desire and acceptance of group-based hierarchy and the domination of "inferior" groups by "superior" groups.[87] One reason for these results could be that Social Dominance Orientation correlates with acceptance of human dominance over nature and animals, lower empathy and expectations of inevitable hierarchies between humans.[88] It has been proposed that such a hierarchical mindset may afford individuals to exploit nature and keep demanding more evidence for climate change before admitting it, as they may be less concerned for those who risk its most negative impacts: disadvantaged people, nonhuman animals and future generations.[89]

Most research regarding polarization of the climate issue among the public have focused on supporters of traditional and mainstream parties but, as described in Chapter 4, populist[90] and far-right sentiments have re-emerged across the Western world over the past decades and changed the political landscapes and influenced environmental discussions.[91] As described in Chapter 4, far-right supporters express, on average, more dismissive attitudes towards climate change than supporters of other parties.[92] Research investigating this link is still scarce, and has particularly focused on the roles of conservative ideologies, exclusionary sociocultural attitudes (opposition to e.g., immigration, feminism and societal focus on minority rights), and populism and institutional distrust. A cumulative body of evidence shows, both among the broader audience and among the populist and far-right segments of the public, that climate obstruction is linked to racial attitudes, opposition to immigration, anti-feminist views, and other forms of exclusionary attitudes.[93] However, as also mentioned in Chapter 4, the roles of populism and distrust need to be further examined because results have been inconsistent.[94] Moreover, a study in Finland found that right-wing populists' rhetoric includes scepticism towards climate change and the so-called soft sciences but no other forms of anti-science attitudes.[95] When these results are interpreted in relation to the observations that climate obstruction is not linked to left-wing populist parties,[96] it seems that the explanation is on the core ideology of these parties, or a mixture of these ideologies and anti-establishment views.[97] It is possible that anti-establishment views are so widespread across political spectrums in the public, that they do not offer clear explanations in and of themselves.[98]

Summary

This chapter aimed to explain the public perceptions and behaviours in relation with the ongoing climate crisis. We began by looking into the gap in attitudes and behaviours, with the aim to understand why beliefs and concern over climate change so seldomly translate to effective climate engagement. We then discussed some reasons why the views on climate science differ among the public, with a focus on basic human information processing and on motivational reasoning that can hamper goals to draw accurate conclusions. And lastly, we discussed psychological factors and processes that could explain why climate obstruction correlates with certain policy preferences, ideologies and identities. Throughout the chapter, we aimed to understand the interplay of individuals and the system, at all levels of obstruction: primary obstruction in interaction with denial machine and disinformation; secondary in relation to political discussions and identity processes; and tertiary as a manifestation of societal structures, norms and ideas of a good life. As such, we highlighted the embeddedness of individual opinions and actions in a web of socio-economic relationships, information environment and societal structures. Together, the results show that, to explain individuals' climate engagement, there is a need to take various factors into account.

With this chapter, we have finished our examinations into different forms of climate obstruction. Next, we integrate our findings and aim to identify the unique and common themes across our fields, and discuss some potential solutions that could help in solving climate crisis.

Notes

1 Eurobarometer (2021) *Special Eurobarometer 513. Climate Change.* Available at: https://ec.europa.eu/clima/system/files/2021-07/report_2021_en.pdf [Last accessed May 10, 2022].

2 Lorenzoni, I., Nicholson-Cole, S. and Whitmarsh, L. (2007) "Barriers perceived to engaging with climate change among the UK public and their policy implications", *Global Environmental Change*, 17, pp.445–459. https://doi.org/10.1016/j.gloenvcha.2007.01.004; Ockwell, D., Whitmarsh, L. and O'Neill, S. (2009) "Reorienting climate change communication for effective mitigation", *Science Communication*, 30(3), pp.305–327. https://doi.org/10.1177/1075547008328969.

3 Lorenzoni, I., Nicholson-Cole, S. and Whitmarsh, L. (2007) "Barriers perceived to engaging with climate change among the UK public and their policy implications"; Ockwell, D., Whitmarsh, L. and O'Neill, S. (2009) "Reorienting climate change communication for effective mitigation"; Nielsen, K.S. *et al.* (2020) "How psychology can help limit climate change", *American Psychologist*, 76(1), pp.130–144. https://doi.org/10.1037/amp0000624; Nielsen, K.S. *et al.* (2021) "The case for impact-focused environmental psychology", *Journal of Environmental Psychology*, 74, 101559. https://doi.org/10.1016/j.jenvp.2021.101559.

4 Ockwell, D., Whitmarsh, L. and O'Neill, S. (2009) "Reorienting climate change communication for effective mitigation".

5 Climate Outreach (2021) "Linking individual action and system change in climate advocacy", *Climate Engagement Lab.*

6 Heglar, Mary Annaise "We cant tackle climate change without you", *Wired.* https://www.wired.com/story/what-you-can-do-solve-climate-change/ [Last accessed May 10, 2022].

7 Fairbrother, M. (2022) "Public opinion about climate policies: a review and call for more studies of what people want", *PLOS Climate*, 1(5), e0000030. https://doi.org/10.1371/journal.pclm.0000030.

8 Hornsey, M.J. and Fielding, K.S. (2017) "Attitude roots and Jiu Jitsu persuasion: understanding and overcoming the motivated rejection of science", *American Psychologist*, 72(5), pp.459–473. https://doi.org/10.1037/a0040437; Jacquet, J., Dietrich, M. and Jost, J.T. (2015) "The ideological divide and climate change opinion: 'top-down' and 'bottom-up' approaches", *Frontiers in Psychology*, 5(1458). https://doi.org/10.3389/fpsyg.2014.01458.

9 Social psychology studies how thoughts, feelings, and behaviours of individuals are influenced by the presence of others and/or the internalised social norms.

10 Maiella, R., La Malva, P., Marchetti, D., Pomarico, E., Di Crosta, A., Palumbo, R., Cetara, L., Di Domenico, A. and Verrocchio, M.C. (2020) "The psychological distance and climate change: a systematic review on the mitigation and adaptation behaviors", *Frontiers in Psychology*, 11, 568899. https://doi.org/10.3389/fpsyg.2020.568899; Spence, A., Poortinga, W. and Pidgeon, N. (2012) "The psychological distance of climate change", *Risk Analysis*, 32, pp.957–972. https://doi.org/10.1111/j.1539-6924.2011.01695.x.

11 Poortinga, W. *et al.* (2011) "Uncertain climate: an investigation into public scepticism about anthropogenic climate change", *Global Environmental Change*, 21, pp.1015–1024.

12 Geiger, N. and Swim, J. (2016) "Climate of silence: pluralistic ignorance as a barrier to climate change discussion", *Journal of Environmental Psychology*, 47, pp.79–90. https://doi.org/10.1016/j.jenvp.2016.05.002.

13 Eurobarometer (2021) *Special Eurobarometer 513. Climate Change.*

14 Leiserowitz, A. *et al.* (2022) *Global Warming's Six Americas, September 2021.* New Haven, CT: Yale Program on Climate Change Communication, Yale University and George Mason University.

15 Tam, K.-P. and Milfont, T.L. (2020) "Towards cross-cultural environmental psychology: a state-of-the-art review and recommendations", *Journal of Environmental Psychology*, 71, 101474. https://doi.org/10.1016/j.jenvp.2020.101474.

16 The G20 Peoples' Climate Vote 2021 United Nations Development Programme. Available at: https://www.undp.org/sites/g/files/zskgke326/files/publications/UNDP-Oxford-Peoples-Climate-Vote-Results.pdf [Last accessed May 13, 2022].

17 Kollmuss, A. and Agyeman, J. (2002) "Mind the gap: why do people act environmentally and what are the barriers to pro-environmental behavior", *Environmental Education Research*, 8, pp.239–260. https://doi.org/10.1080/13504620220145401.

18 American Psychological Association, APA Task Force on Climate Change (2022) "Addressing the climate crisis: An action plan for psychologists, report of the APA task force on climate change". https://www.apa.org/science/about/publications/climate-crisis-action-plan.pdf [Last accessed May 12, 2022]; Hines, J.M., Hungerford, H.R. and Tomera, A.N. (1986) "Analysis and synthesis of research on responsible pro-environmental behavior: a meta-analysis", *The Journal of Environmental Education*, 18(2), pp.1–8. https://doi.org/10.1080/00958964.1987.9943482; Banberg, S. and Möser, G. (2007) "Twenty years after Hines, Hungerford, and Tomera: a new meta-analysis of psychosocial determinants of pro-environmental behaviour", *Journal of Environmental Psychology*, 27, pp.14–25. https://doi.org/10.1016/j.jenvp.2006.12.002; Blake, J. (1999) "Overcoming the 'value–action gap' in environmental policy: tensions between national policy and local experience", *Local Environment*, 4(3), pp.257–278. https://doi.org/10.1080/13549839908725599; Kollmuss, A. and Agyeman, J. (2002) "Mind the gap".

19 See for example Steg, L. and Vlek, C. (2009) "Encouraging pro-environmental behaviour: an integrative review and research agenda", *Journal of Environmental Psychology*, 29, pp.309–317. https://doi.org/10.1016/j.jenvp.2008.10.004.

20 Kaiser, F.G., Byrka, K. and Hartig, T. (2010) "Reviving Campbell's Paradigm for attitude research", *Personality and Social Psychology Review*, 14(4), pp.351–367. https://doi.org/10.1177/1088868310366452; Vilar, R., Milfont, T.L. and Sibley, C.G. (2020) "The role of social desirability responding in the longitudinal relations between intention and behaviour", *Journal of Environmental Psychology*, 70, 101457. https://doi.org/10.1016/j.jenvp.2020.101457.

21 Kaiser, F.G., Byrka, K. and Hartig, T. (2010) "Reviving Campbell's Paradigm for attitude research".

22 Eurobarometer (2021) *Special Eurobarometer 513. Climate Change.*

23 Ibid; Macdiarmid, J.I., Douglas, F. and Campbel, J. (2016) "Eating like there's no tomorrow: public awareness of the environmental impact of food and reluctance to eat less meat as part of a sustainable diet", *Appetite*, 96, pp.486–493. https://doi.org/10.1016/j.appet.2015.10.011.

24 Kollmuss, A. and Agyeman, J. (2002) "Mind the gap"; Nielsen, K.S. *et al.* (2021) "The case for impact-focused environmental psychology".

25 Wynes, S. and Nicholas, K.A. (2017) "The climate mitigation gap: education and government recommendations miss the most effective individual actions", *Environmental Research Letters*, 12, 074024. https://doi.org/10.1088/1748-9326/aa7541; Macdiarmid, J.I., Douglas, F. and Campbel, J. (2016) "Eating like there's no tomorrow".

26 Gössling, S. (2019) "Celebrities, air travel, and social norms", *Annals of Tourism Research*, 79, 102775. https://doi.org/10.1016/j.annals.2019. 102775; Steg, L. and Vlek, C. (2009) "Encouraging pro-environmental behaviour"; Whitmarsh, L. and O'Neill, S. (2010) "Green identity, green living? The role of pro-environmental self-identity in determining consistency across diverse pro-environmental behaviours", *Journal of Environmental Psychology*, 30(3), pp.305–314. https://doi.org/10.1016/j. jenvp.2010.01.003; Nielsen, K.S. *et al.* (2021) "The case for impact-focused environmental psychology".

27 Lorenzoni, I., Nicholson-Cole, S. and Whitmarsh, L. (2007) "Barriers perceived to engaging with climate change among the UK public and their policy implications".

28 Ibid; Ockwell, D., Whitmarsh, L. and O'Neill, S. (2009) "Reorienting climate change communication for effective mitigation".

29 Hyry, J. (2021) "Climate emotions: Summary of key findings", *SITRA*. Available at: https://media.sitra.fi/2019/11/29131052/sitraclimate-emotions-report-2019.pdf [Last accessed May 10, 2022].

30 Oxfam (2020) "Confronting Carbon Inequality: Putting Climate Justice at the Heart of the COVID-19 Recovery", Oxfam Media briefing, p.3. Available at: https://oxfamilibrary.openrepository.com/bitstream/handle/ 10546/621052/mb-confronting-carbon-inequality-210920-en.pdf [Last accessed May 12, 2022]. The image is reproduced with the permission of Oxfam, Oxfam House, John Smith Drive, Cowley, Oxford OX4 2JY, UK. www.oxfam.org.uk. Oxfam does not necessarily endorse any text or activities that accompany the materials.

31 Szeman, I. (2019) *On Petrocultures: Globalization, Culture, and Energy.* West Virginia University Press.

32 Macdiarmid, J.I., Douglas, F. and Campbel, J. (2016) "Eating like there's no tomorrow".

33 Dhont, K. and Hodson, G. (2014) "Why do right-wing adherents engage in more animal exploitation and meat consumption?", *Personality and Individual Differences*, 64, pp.12–17. https://doi.org/10.1016/J.PAID.2014. 02.002.

34 Aavik, K. (2021) "Vegan men: towards greater care for (non) human others, earth and self" in Pulé, P. and Hultman, M. (eds) *Men, Masculinities, and Earth*. Cham: Palgrave Macmillan, pp.329–350.

35 Norgaard, K.M. (2011) *Living in Denial: Climate Change, Emotions, and Everyday Life*. Cambridge: MIT Press.

36 Ibid., pp.43–44.

37 Ibid.
38 Festinger, L. (1957) *A Theory of Cognitive Dissonance*. Standford, CA: Standford University Press.
39 Wullenkord, M.C. and Reese, G. (2021) "Avoidance, rationalization, and denial: defensive self-protection in the face of climate change negatively predicts pro-environmental behavior", *Journal of Environmental Psychology*, 77, 101683. https://doi.org/10.1016/j.jenvp.2021.101683.
40 Bandura, A. (1977) "Self-efficacy: toward a unifying theory of behavioral change", *Psychological Review*, 84(2), p.191; Ojala, M. (2012) "Regulating worry, promoting hope: how do children, adolescents, and young adults cope with climate change?", *International Journal of Environmental & Science Education*, 7(4), pp.537–561; Ojala, M. (2015) "Climate change skepticism among adolescents", *Journal of Youth Studies*, 18(9), pp.1135–1153. https://doi.org/10.1080/13676261.2015.1020927; Ojala, M., Cunsolo, A., Ogunbode, C.A. and Middleton, J. (2021) "Anxiety, worry, and grief in a time of environmental and climate crisis: a narrative review", *Annual Review of Environment and Resources*, 46, pp.35–58. https://doi.org/10.1146/annurev-environ-012220-022716.
41 Hornsey, M.J. and Fielding, K.S. (2016) "Cautionary note about messages of hope: focusing on progress in reducing carbon emissions weakens mitigation motivation", *Global Environmental Change*, 39, pp. 26–34. https://doi.org/10.1016/j.gloenvcha.2016.04.003.
42 Bury, S.M., Wenzel, M. and Woodyatt, L. (2020) "Against the odds: hope as an antecedent of support for climate change action", *British Journal of Social Psychology*, 59(2), pp.289–310. https://doi.org/10.1111/bjso.12343; Marlon, J.R. *et al.* (2019) "How hope and doubt affect climate change mobilization", *Frontiers in Communication*, 4. https://doi.org/10.3389/fcomm.2019.00020; Ojala, M. (2012) "Regulating worry, promoting hope"; Ojala, M., Cunsolo, A., Ogunbode, C.A. and Middleton, J. (2021) "Anxiety, worry, and grief in a time of environmental and climate crisis: a narrative review". https://doi.org/10.1146/annurev-environ-012220-022716.
43 Oscarsson, H., Strömbäck, J. and Jönsson, E. (2021) "Svenska klimatförnekare" in Andersson, U., Carlander, A., Grusell, M. and Öhberg, P. (eds) *Ingen anledning till oro (?)*. Gothenburg: SOM-Institute, University of Gothenburg.
44 Burck, J. *et al.* (2021) "Climate Change Performance Index 2022", Germanwatch, NewClimate Institute & Climate Action Network. Available at: https://ccpi.org/download/climate-change-performance-index-2022-2/ [Last accessed May 11, 2022].
45 Oscarsson, H., Strömbäck, J. and Jönsson, E. (2021) "Svenska klimatförnekare".
46 Leviston, Z. and Walker, I. (2012) "Beliefs and denials about climate change: an Australian perspective", *Ecopsychology*, 4, pp.277–285. https://doi.org/10.1089/eco.2012.0051; Poortinga, W. *et al.* (2011) "Uncertain climate: an investigation into public scepticism about anthropogenic climate change", *Global Environmental Change*, 21, pp.1015–1024. https://doi.org/10.1016/j.gloenvcha.2011.03.001.
47 Tversky, A. and Kahneman, D. (1974) "Judgment under uncertainty: heuristics and biases", *Science*, 185, pp.1124–1131.

48 Heider, F. (1958) *The Psychology of Interpersonal Relations*. New York: Wiley.

49 Hendriks, F., Kienhues, D. and Bromme, R. (2016) "Trust in science and the science of trust", in B. Blöbaum (ed), *Trust and Communication in a Digitized World: Models and Concepts of Trust Research*. Springer International Publishing, pp.143–159. https://doi.org/10.1007/978-3-319-28059-2_8

50 Diethelm, P. and McKee, M. (2009) "Denialism: what is it and how should scientists respond?", *The European Journal of Public Health*, 19(1), pp.2–4. https://doi.org/10.1093/eurpub/ckn139; Oreskes, N. and Conway, E.M. (2011) *Merchants of Doubt: How a Handful of Scientists Obscured the Truth on Issues from Tobacco Smoke to Global Warming*. Bloomsbury Publishing USA.

51 Budescu, D.V., Broomell, S.B. and Por, H. (2009) "Improving communication of uncertainty in the reports of the intergovernmental panel on climate change", *Psychological Science*, 20(3), pp.299–308. https://doi.org/10.1111/j.1467-9280.2009.02284.x.

52 Oreskes, N. and Conway, E.M. (2011) *Merchants of Doubt*; Pasek, A. (2021) "Carbon vitalism: life and the body in climate denial", *Environmental Humanities*, 13(1), pp.1–20. https://doi.org/10.1215/22011919-8867175.

53 Lewandowsky, S., Mann, M.E., Brown, N.J.L. and Friedman, H. (2016) "Science and the public: debate, denial, and skepticism", *Journal of Social and Political Psychology*, 4(2), pp.537–553. https://doi.org/10.5964/jspp. v4i2.604; Normand, M.P. (2008) "Science, skepticism, and applied behavior analysis", *Behavior Analysis in Practice*, 1(2), pp.42–49. https://doi. org/10.1007/BF03391727.

54 See for example Lewandowsky, S., Mann, M.E., Brown, N.J.L. and Friedman, H. (2016) "Science and the public: debate, denial, and skepticism"; Oreskes, N. and Conway, E.M. (2011) *Merchants of Doubt*; Washington, H. and Cook, J. (2011) *Climate Change Denial: Heads in the Sand*. London, UK: Earthscan.

55 Jylhä, K.M. (2018) "Denial versus reality of climate change" in DellaSala, D.A. and Goldstein, M.I. (eds) *Encyclopedia of the Anthropocene*. Oxford: Elsevier, pp.487–492. https://doi.org/10.1016/B978-0-12-809665-9.09762-7.

56 Cook, J. *et al.* (2019) "America misled: how the fossil fuel industry deliberately misled Americans about climate change", Fairfax, VA: George Mason University Center for Climate Change Communication, p.6. Available at: https://www.climatechangecommunication.org/america-misled/ [Last accessed March 29, 2022].

57 Kunda, Z. (1990) "The case for motivated reasoning", *Psychological Bulletin*, 108, pp.480–498.

58 Nickerson, R.S. (1998) "Confirmation bias: a ubiquitous phenomenon in many guises", *Review of General Psychology*, 2(2), pp.75–220.

59 Campbell, T.H. and Kay, A.C. (2014) "Solution aversion: on the relation between ideology and motivated disbelief", *Journal of Personality and Social Psychology*, 107(5), pp.809–824. https://doi.org/10.1037/a0037963.

60 Franta, B. (2021) "Weaponizing economics: big oil, economic consultants, and climate policy delay", *Environmental Politics*. https://doi.org/10.1080/09644016.2021.1947636.

61 See for example Hornsey, M.J., Harris, E.A., Bain, P.G. and Fielding, K.S. (2016) "Meta-analyses of the determinants and outcomes of belief in

climate change"; Hornsey, M.J., Harris, E.A. and Fielding, K.S. (2018) "Relationships among conspiratorial beliefs, conservatism and climate scepticism across nations", *Nature Climate Change*, 8, pp.614–620. https://doi.org/10.1038/s41558-018-0157-2.

62 Hornsey, M.J., Harris, E.A., Bain, P.G. and Fielding, K.S. (2016) "Meta-analyses of the determinants and outcomes of belief in climate change".

63 Jylhä, K.M., Cantal, C., Akrami, N. and Milfont, T.L. (2016) "Denial of anthropogenic climate change: social dominance orientation helps explain the conservative male effect in Brazil and Sweden", *Personality and Individual Differences*, 98, pp.184–187. https://doi.org/10.1016/j.paid.2016.04.020; McCright, A.M., Dunlap, R.E. and Marquart-Pyatt, S.T. (2016) "Political ideology and views about climate change in the European Union", *Environmental Politics*, 25, pp.338–358. https://doi.org/10.1080/09644016.2015.1090371.

64 Carrus, G., Panno, A. and Leone, L. (2018) "The moderating role of interest in politics on the relations between conservative political orientation and denial of climate change", *Society & Natural Resources*, 31(10), pp.1103–1117. https://doi.org/10.1080/08941920.2018.1463422.

65 Harring, M. and Sohlberg, J. (2017) "The varying effects of left–right ideology on support for the environment: evidence from a Swedish survey experiment", *Environmental Politics*, 26(2), pp.278–300. https://doi.org/10.1080/09644016.2016.1244965.

66 McCright, A.M. and Dunlap, R.E. (2003) "Defeating Kyoto: the conservative movement's impact on US climate change policy", *Social Problems*, 50(3), pp.348–373. https://doi.org/10.1525/sp.2003.50.3.348; Whitmarsh, L. and Corner, A. (2017) "Tools for a new climate conversation".

67 Jacques, P.J., Dunlap, R.E. and Freeman, M. (2008) "The organisation of denial: conservative think tanks and environmental skepticism", *Environmental Politics*, 17(3), pp.349–385. https://doi.org/10.1080/09644010802055576.

68 Van Bavel, J.J. and Pereira, A. (2018) "The partisan brain: an identity-based model of political belief", *Trends in Cognitive Sciences*, 22(3), pp.201–224. https://doi.org/10.1016/j.tics.2018.01.004.

69 Lönnqvist, J.-E., Ilmarinen, V.-J. and Sortheix, F.M. (2020) "Polarization in the wake of the European refugee crisis – a longitudinal study of the Finnish political elite's attitudes towards refugees and the environment", *Journal of Social and Political Psychology*, 8(1), pp.173–197. https://doi.org/10.5964/jspp.v8i1.1236.

70 Ibid.

71 Clarke, E.J. *et al.* (2019) "Mitigation system threat partially mediates the effects of right-wing ideologies on climate change beliefs", *Journal of Applied Social Psychology*, 49(6), pp.349–360. https://doi.org/10.1111/jasp.12585; Doell, K.C. *et al.* (2021) "Understanding the effects of partisan identity on climate change", *Current Opinion in Behavioral Sciences*, 42, pp.54–59. https://doi.org/10.1016/j.cobeha.2021.03.013; Hoffarth, M.R. and Hodson, G. (2016) "Green on the outside, red on the inside: perceived environmentalist threat as a factor explaining political polarization of climate change", *Journal of Environmental Psychology*, 45, pp.40–49. https://doi.org/10.1016/j.jenvp.2015.11.002.

72 Milfont, T.L. and Sibley, C.G. (2016) "Empathic and social dominance orientations help explain gender differences in environmentalism: a one-year Bayesian mediation analysis", *Personality and Individual Differences*, 90, pp.85–88. https://doi.org/10.1016/j.paid.2015.10.044; Zelezny, L.C., Chua, P. and Aldrich, C. (2000) "Elaborating on gender differences in environmentalism", *Journal of Social Issues. Special Issue: Promoting Environmentalism*, 56, pp.443–457. https://doi.org/10.1111/0022-4537.00177.

73 See for example Jylhä, K.M., Strimling, P. and Rydgren, J. (2020) "Climate change denial among radical right-wing supporters", *Sustainability*, 12, 10226. https://doi.org/10.3390/su122310226; Nicol, A.A.M., De France, K. and Nicol, A.M. (2022) "The relation of climate change denial with benevolent and hostile sexism", *Journal of Applied Social Psychology*, 1–14. https://doi.org/10.1111/jasp.12880; Salmen, A. and Dhont, K. (2020) "Hostile and benevolent sexism: the differential roles of human supremacy beliefs, women's connection to nature, and the dehumanization of women", *Group Processes & Intergroup Relations*, 24(7), pp.1053–1076. https://doi.org/10.1177/1368430220920713.

74 Pulé, P.M. and Hultman, M. (eds) (2021) *Men, Masculinities, and Earth: Contending with the (m)Anthropocene*. London: Palgrave Macmillan.

75 Jacquet, J., Dietrich, M. and Jost, J.T. (2015) "The ideological divide and climate change opinion".

76 Bakker, B.N., Rooduijn, M. and Schumacher, G. (2016) "The psychological roots of populist voting: evidence from the United States, the Netherlands and Germany", *European Journal of Political Research*, 55, pp.302–320. https://doi.org/10.1111/1475-6765.12121; Caprara, G.V. and Zimbardo, P.G. (2004) "Personalizing politics: a congruency model of political preference", *American Psychologist*, 59(7), pp.582–594. https://doi.org/10.1037/0003-066X.59.7.581.

77 Harteveld, E., Kokkonen, A. and Dahlberg, S. (2017) "Adapting to party lines: the effect of party affiliation on immigration attitudes", *West European Politics*, 40(6), pp.1177–1197. https://doi.org/10.1080/01402382.2017.1328889.

78 Campbell, T.H. and Kay, A.C. (2014) "Solution aversion".

79 Azevedo, F. and Jost, J.T. (2021) "The ideological basis of antiscientific attitudes: effects of authoritarianism, conservatism, religiosity, social dominance, and system justification", *Group Processes & Intergroup Relations*, 24(4), pp.518–549. https://doi.org/10.1177/1368430221990104;- Jylhä, K.M., Strimling, P. and Rydgren, J. (2020) "Climate change denial among radical right-wing supporters".

80 Harring, M. and Sohlberg, J. (2017) "The varying effects of left–right ideology on support for the environment: evidence from a Swedish survey experiment", *Environmental Politics*, 26(2), pp.278–300. https://doi.org/10.1080/09644016.2016.1244965.

81 Feygina, I., Jost, J.T. and Goldsmith, R.E. (2010) "System justification, the denial of global warming, and the possibility of 'system-sanctioned change'", *Personality and Social Psychology Bulleting*, 36(3), pp.326–338. https://doi.org/10.1177/0146167209351435.

82 Milfont, T.L. and Sibley, C.G. (2014) "The hierarchy enforcement hypothesis of environmental exploitation: a social dominance perspective", *Journal of Experimental Social Psychology*, 55, pp.188–193. https://doi.org/10.1016/j.jesp.2014.07.006.

83 Jost, J.T. and Banaji, M.R. (1994) "The role of stereotyping in system-justification and the production of false consciousness", *British Journal of Social Psychology*,33(1),pp.1–27.https://doi.org/10.1111/j.2044-8309.1994.tb01008.x

84 Feygina, I., Jost, J.T. and Goldsmith, R.E. (2010) "System justification, the denial of global warming, and the possibility of 'system-sanctioned change'".

85 Such a link has been found with System Justification (motivation to see the overarching societal system as fair, desirable and natural) and Right-Wing Authoritarianism (an attitudinal cluster including readiness to accept and follow established authorities, aggression towards those whose lifestyle and values threatens the social order, and preference and commitment to traditional norms and lifestyles), see, for example, Feygina, I., Jost, J.T. and Goldsmith, R.E. (2010) "System justification, the denial of global warming, and the possibility of 'system-sanctioned change'"; Jylhä, K.M. and Akrami, N. (2015) "Social dominance orientation and climate change denial: the role of dominance and system justification", *Personality and Individual Differences*, 86, pp.108–111. https://doi.org/10.1016/j.paid.2015.05.041; Schultz, P.W. and Stone, W.F. (1994) "Authoritarianism and attitudes toward the environment", Stanley, S.K. and Wilson, M.S. (2019) "Meta-analysing the association between social dominance orientation, authoritarianism, and attitudes on the environment and climate change", *Journal of Environmental Psychology*, 61, pp.46–56. https://doi.org/10.1016/j.jenvp.2018.12.002.

86 Stanley, S.K. and Wilson, M.S. (2019) "Meta-analysing the association between social dominance orientation, authoritarianism, and attitudes on the environment and climate change", *Journal of Environmental Psychology*, 61, pp.46–56. https://doi.org/10.1016/j.jenvp.2018.12.002.

87 Pratto, F., Sidanius, J., Stallworth, L.M. and Malle, B.F. (1994) "Social dominance orientation: a personality variable predicting social and political attitudes", *Journal of Personality and Social Psychology*, 67(4), pp.741–763. https://doi.org/10.1037/0022-3514.67.4.741.

88 Jylhä, K.M. and Akrami, N. (2015) "Social dominance orientation and climate change denial: the role of dominance and system justification"; Jylhä, K.M., Tam, K.-P. and Milfont, T.L. (2020) "Acceptance of group-based dominance and climate change denial: a cross-cultural study in Hong Kong, New Zealand, and Sweden", *Asian Journal of Social Psychology*, 24(2), pp.198–207. https://doi.org/10.1111/ajsp.12444; Milfont, T.L. *et al.* (2013) "Environmental consequences of the desire to dominate and be superior", *Personality and Social Psychology Bulletin*, 39, pp.1127–1138. https://doi.org/10.1177/0146167213490805.

89 Jackson, L.M., Bitacola, L.M., Janes, L.M. and Esses, V.M. (2013) "Intergroup ideology and environmental inequality", *Analyses of Social Issues and Public Policy*, 13, pp.327–346. https://doi.org/10.1111/asap.12035; Jylhä, K.M., Cantal, C., Akrami, N. and Milfont, T.L. (2016) "Denial of

anthropogenic climate change: social dominance orientation helps explain the conservative male effect in Brazil and Sweden".

90 Various definitions to populism exist but it can be defined as an ideology (or rhetorical style) that divides society into two heterogenous groups: the immoral and malevolent political 'elites' and the virtuous and sensible 'people' who are portrayed as the victims of the elite, see for example Mudde, C. (2004) "The populist zeitgeist", *Government and Opposition*, 39, pp.541–563. https://doi.org/10.1111/j.1477-7053.2004.00135.x.

91 Rydgren, J. (2007) "The sociology of the radical right", *Annual Review of Sociology*, 33, pp.241–262.

92 Huber, R.A. (2020) "The role of populist attitudes in explaining climate change skepticism and support for environmental protection"; Jylhä, K.M., Strimling, P. and Rydgren, J. (2020) "Climate change denial among radical right-wing supporters".

93 Benegal, S.D. (2018) "The spillover of race and racial attitudes into public opinion about climate change", *Environmental Politics*, 27(4), pp.733–756. https://doi.org/10.1080/09644016.2018.1457287; Bloodheart, B. and Swim, J. (2010) "Equality, harmony, and the environment: an ecofeminist approach to understanding the role of cultural values on the treatment of women and nature", *Ecopsychology*, 2(3), pp.187–194. https://doi.org/10.1089/eco.2010.0057; Jylhä, K.M. and Hellmer, K. (2020) "Right-wing populism and climate change denial"; Jylhä, K.M., Strimling, P. and Rydgren, J. (2020) "Climate change denial among radical right-wing supporters"; Krange, O., Kaltenborn, B.P. and Hultman, M. (2018) "Cool dudes in Norway: climate change denial among conservative Norwegian men", *Environmental Sociology*, 5(1), pp.1–11. https://doi.org/10.1080/2325 1042.2018.1488516; Nicol, A.A.M, De France, K. and Nicol, A.M. (2022) "The relation of climate change denial with benevolent and hostile sexism"; Ojala, M. (2015) "Climate change skepticism among adolescents".

94 See for example Huber, R.A. (2020) "The role of populist attitudes in explaining climate change skepticism and support for environmental protection"; Jylhä, K.M. and Hellmer, K. (2020) "Right-wing populism and climate change denial".

95 Ylä-Anttila, T. (2018) "Populist knowledge: 'post-truth' repertoires of contesting epistemic authorities", *European Journal of Cultural and Political Sociology*, 5(4), pp.356–388. https://doi.org/10.1080/23254823.20 17.1414620.

96 Yan, P., Schroeder, R. and Stier, S. (2021) "Is there a link between climate change scepticism and populism? An analysis of web tracking and survey data from Europe and the US", *Information, Communication & Society*. https://doi.org/10.1080/1369118X.2020.1864005.

97 Jylhä, K.M., Strimling, P. and Rydgren, J. (2020) "Climate change denial among radical right-wing supporters"; Lockwood, M. (2018) "Right-wing populism and the climate change agenda: exploring the linkages".

98 Jylhä, K.M. and Hellmer, K. (2020) "Right-wing populism and climate change denial"; For example, 54 per cent of Europeans think the interests of people like them are not properly taken into account by the political system in their country: Special Eurobarometer 486. Available at: http://data. europa.eu/88u/dataset/s2225_91_2_486_eng [Last accessed May 11, 2022].

6 Conclusion

Introduction

As climate science has become more robust over the past few decades, there is no longer any doubt that the burning of fossil fuels is altering the very circumstances for life on Earth. Among climate scientists, this conclusion has long been uncontroversial[1] and was yet again stated by the Working Group I of the Sixth Assessment Report of IPCC in 2021: "[i]t is unequivocal that human influence has warmed the atmosphere, ocean and land. Widespread and rapid changes in the atmosphere, ocean, cryosphere and biosphere have occurred".[2] In this book, we have described how this knowledge and the adherent necessary actions have been obstructed over decades at several levels of society. Drawing on an ever-growing number of academic studies, we have pointed to the organisation and affinities of those who have been most forcefully blocking serious action on climate change, particularly since the establishment of the IPCC in 1988. We highlighted the role of industry, especially the fossil fuel industry, as well as political sectors – including the comparatively recent role played by the far right. We have also shed light on societal structures which make it difficult to mobilise popular engagement and support for effective climate solutions. Together, these dynamics hamper the efforts to end the burning of fossil fuels and to adapt and protect against the already occurring changes.

As is evident, obstacles standing in the way of effective climate policy are not limited to outright denial of anthropogenic climate change, or the lobbying by particular actors with vested interests, but include a wide range of strategies and tactics used and maintained across society. To cover this spectrum, we have drawn on the concept of obstruction as an umbrella term to provide a novel, comprehensive framework of the different forms (primary, secondary and tertiary) of climate

DOI: 10.4324/9781003181132-6

obstruction and their interactions, as well as their unique and shared manifestations and foundations.

Importantly, the concept of obstruction highlights the agency in processes but also the unintended behaviours and the consequences of (non-)actions. Current lifestyles and societal structures in many parts of society form a blockage for reaching meaningful climate goals. This blockage is maintained through corporate efforts, political inaction and infrastructural lock-ins. However, it is also sustained through the countless everyday actions that reproduce an unsustainable system, including the repeated decisions to *not* act and the behaviours that people are sometimes forced to choose in the absence of relevant/necessary infrastructure. While we claim in accordance with recent scholarship that obstruction provides a way to grasp the many ways climate change action is hindered, we have also found it necessary to differentiate the term to shed light on the complexity of the issue.[3] That is, we have introduced obstruction as a three-layered, intertwined activity, unfolding with regard to science and policy and across the economic, political and cultural spheres.

Speaking of primary, secondary and tertiary obstruction, the former describes the wilful or ignorant dismissals of science and opposition to mitigation policies. Primary obstruction has been framed as denial (or scepticism) and is carried out most forcefully by the fossil fuel industry and business interest groups. Denial is also still spreading among the public and a sizeable minority still doubts, or even outright rejects, the established scientific findings on anthropogenic climate change and oppose possible solutions.[4] Secondary obstruction denotes all those calls to forestall action in the name of (alleged) uncertainty or speculated costs of action. Here too, corporate actors are crucial, but political actors have also long played a key role. Moreover, there is disagreement among the public regarding which solutions are the most preferable ones, and some policies and actions are met with low support or even opposition, such as is the case with carbon taxes or plant-based diets.[5] Last, tertiary obstruction is comprised of all those activities in which even well-meaning individuals and societies engage in, and do not want to or cannot take the steps needed to curb emissions. This third layer thus highlights the complex embeddedness of individual lifestyle choices in the broader system by incorporating the individual-level explanations and the cultural and structural lock-ins that have been created during the last 150 years. As such, we show that each of these three forms of obstruction relates to one another, though they differ in terms of responsibility and power to influence.

Commonalities in the landscape of obstruction

What comes to light when combining knowledge from our different perspectives and fields of expertise? First, one key aspect permeating all our fields is the relevance and importance of dominance and protection of hierarchies in creating and shaping climate change obstruction. As such, we can see that opposition to environmental protection and the relativisation of climate science is forcefully orchestrated by major fossil fuel players and industry leaders. As shown throughout this book, political ideologies and agendas that build on conservatism and protecting ethnic/racial or class-based (real or imagined) hierarchies are the ones consistently linked to obstructing climate science and action. Similarly, capitalist relations permeating societies have enabled huge wealth to be accumulated with certain individuals, companies and states. For example, even though obstruction of climate policies largely emerged as a phenomenon in the 1990s, the aim to protect Western dominance sparked the neoliberal backlash against environmentalism and environmental regulation as early as the 1970s. Today, far-right actors are often vocal in their opposition to climate protection when claiming to defend jobs, infrastructures and lifestyles at the expense of the environment and "others". Similar tendencies are also shown in psychological research, as various forms of obstruction correlate with the acceptance and promotion of social group-based dominance and hierarchies, and human dominance over nature and animals.[6]

The call to end business-as-usual thus entails more than just ending, or at least heavily restricting, the production of fossil fuels.[7] Ending business as usual also means the end of a system and mode of organising society that have historically built on the exploitation of lands and resources in the Global South for the benefit of the Global North. This environmental injustice is also explicitly shown in the contemporary climate context, in the unevenly distributed risks, burdens and benefits.[8] While the (relatively) wealthy cause climate change, other groups are at greater risk, and also typically have very little influence over the situation: the disadvantaged people and the young have rarely been heard in climate negotiations and politics, and the future generation and non-human animals cannot raise their voices and claim their rights.[9] These conclusions – that to overcome obstruction we also need to undo some deeply rooted material and cultural ways of life in many parts of the world – is something that has been forcefully stated by scholars working on the intersection of climate change and coloniality. Commenting on the famous line: "it is easier to imagine an end to the world than an end to capitalism", Amitav Ghosh writes, "That which is really harder to imagine than the end of the world is the end of the absolute geopolitical dominance of the West".[10] This speaks to the moment in the 1960s and 1970s when this order was threatened by the conjoined forces of environmentalism, workers movements and liberation struggles in former colonies which, as we have seen, led to a forceful backlash.

Focusing on the colonial aspect of climate obstruction is also important since many of the solutions proposed by countries in the Global North are contingent on the continued use of acreages in the South. One such solution is the Clean Development Mechanism in which a country bound by the Kyoto Protocol can trade territorial emissions, with reductions in developing countries.[11] In other words, what permeates climate obstruction, from the individual to the fossil fuel corporation, is the maintenance of a specific world order that implies the right – if not obligation – to *conquer*, *use* and *extract* from the natural world. While such ideas historically have been most pervasive in a European context, a similar logic exists around the world which also influences obstruction in the Global South.[12]

Second, it is important to underline the flexibility in how climate action is opposed. This is true both for the actors of denial machine and among policymakers and the public. Various arguments and strategies are used, so much so that sometimes views are in conflict with each other. In the U.S. literature, the concept of the playbook is widely used to describe this situation. This metaphor is apt to describe the use of various moves depending on the changing landscape and development of opposition. Importantly, it also directs our attention to how new plays can be added or altered depending on context.

It is therefore crucial to understand obstruction within the specific setting it takes place. Much of the activities in the 1970s and 1980s were aimed at shaping the discourse on environmental problems and forcefully incorporating it into a logic whereby economic calculation and consumer sovereignty became the go-to mode of international environmental governance. This framing has also been pervasive among citizens.[13] In Europe, this framing seems to have remained in place for the most part of the 1990s and early 2000s, all during a period when island nations and the Global South critiqued the insufficient measures taken to protect their existence.[14] In the USA, the strategy was more confrontative and the outright denial of science became a way to stall action which, in turn, watered down the various treaties and declarations that followed. The more confrontative stance is not surprising, given the intimate relation between fossil fuel extraction, the USA and the global financial system after 1971. With the rise of the far right, especially following the 2008 crisis, obstruction has also become part of party-political mobilisation in Europe. Indeed, far-right actors have at times been primary obstructionists, voicing outrageous conspiracy theories. However, and overall, their rhetoric too reflects the wider developments in the corporate sector. That is, far-right climate change communication too is secondary obstruction, informed by ethnonational desires to protect "our" jobs and norms regarding, for example, gender. How these messages have spread and been disseminated is constantly changing but, denial of science, mobilisation of think tanks, lobbying, PR, the use of fake experts and the creation of astro-turf organisations have all been ways to oppose and block serious climate policy through direct opposition but also through shaping the very way we talk about energy, environment and the economy.[15] And while it is thus true that tactics are changing, obstruction has simultaneously existed in its primary, secondary and tertiary forms.

Third and finally, the effectiveness of the denial machine hinges on the ability to strike a chord on the deeply rooted ideas of what constitutes a good life and what kind of energy is needed for this life to remain.[16] Ideas about dominance and the global character of capitalism discussed above thus need to be integrated and normalised in culture for it to become as pervasive as it has become. For example, since the 1960s many societies have built their transportation infrastructure around the expectation that every family owns a car; such ideas are not easily reversed.[17] Some research by social scientists and those working in the humanities have characterised this deep embeddedness of fossil fuels in society as petroculture.[18] Others have argued that energy regimes have shaped our political ideas and societies in even more thorough ways, providing the material foundation for our understanding of concepts such as freedom and democracy.[19] The failure or unwillingness to act at an individual level is therefore understandable, considering our embeddedness in such a material culture. Are we perhaps as societies living in denial?[20]

Indeed, even though the effects of climate change are clearly visible, and one might thus acknowledge the severity of the crisis, sociological research and ethnographical accounts show that obstruction can nevertheless be present.[21] Norgaard and Brulle have described this inaction as a way to avoid cultural trauma resulting from the large socio-technical changes needed to address climate change.[22] Whether we like it or not, much of what many in the world equate with a good life is directly linked to cultures of consumption and high-energy outputs. Similarly, psychological literature has emphasised the rather common human need to feel a sense of certainty, stability, safety and affiliation with other people from their group.[23] Adherence to traditions, ideologies and the overarching societal system can fulfil these needs, which could explain why many people feel threatened by the demands to change the system to, for example, mitigate climate change. Thus, system justifying responses can be evoked, with an aim to provide rationalisation and legitimacy for our current practices.[24]

This is particularly evident in, for example, Sweden where the changes brought by an energy transition since the 1970s have only been carried out in conjunction with the promise to maintain the existing high-energy use way of life.[25] Accordingly, anxieties and fear of loss of privilege or autonomy – often not simply "invented" by corporate and (far-right) political actors but well-founded in the fabric of our societies – do also make climate obstruction at times prominent. Again, what is defended by climate obstruction are existing hierarchical structures and ways of living.[26] It is thus of great importance to acknowledge the role of culture and society in shaping and limiting attitudes and behaviours.

Overcoming obstruction

If forms of obstruction are the problem and current policies are not nearly enough, what are the solutions to counter the ecological breakdown of our planetary boundaries?[27] For too long, environmental governance has relied on the promise of unproven technologies. Yes, technological innovations are valuable, but this should not stop us from acting now.[28] Indeed, responses to the various types of obstruction are manifold, some being more promising in the short term than others. What can, however, serve as a starting point is the insight that solutions which are solely or even primarily directed towards ordinary individuals are not going to be good enough. Rather, individual action or lifestyle changes are intertwined with and unfold within structures.[29]

Confronting climate obstruction means confronting powerful interests in different forms. Industries and the energy sector are not only powerful as they shape laws, investments and subsidies. Rather, fossil fuels are literally powering society and further empowering those with direct control over this resource.[30] To deal with primary obstruction is thus to deal with fossil fuel dependencies permeating relations across various levels, from geopolitics to our private lives. In this sense, the fossil fuel industry is right in pointing to the decisive role their product has for modern life, one they have been complicit in creating.[31] Therefore, the transition to a non-fossil future means structural and comprehensive changes for practically all of us on this planet since oil, coal and gas make up a bit more than 80 per cent of the sold and used energy.[32] Fossil fuels are according to World Trade Organisation as of 2017 subsidised by 6.5 per cent of global GDP.[33] Since the power of fossil fuels is entangled with state investments and subsidies, leaving fossil fuels in the ground is a necessary part of the solution.[34] True, targeting big producers of emissions is vital, but, so is targeting big users of energy on national, company and individual levels. One effective policy change would be to have energy use and emissions accounted for on an individual level, and create specific policies directed towards the super rich who can afford to make large emission cuts.[35] Such "Island Kings" who act as if they are cut off from the rest of our planet have not been called out, identified, circumscribed or dismantled effectively but could through sanctions.[36]

As we have seen throughout the book, the main actors in delaying action are the fossil fuel companies. Through misleading or false narratives, fossil fuel companies have been effective in stalling action for a long time. Exposing their actions and the responsibility they have for the climate crisis has increasingly resulted in legal action against them as well as against states failing to meet climate targets.[37] This approach has been crucial to further reveal the existence and extent of misinformation.[38] Suing fossil fuel companies or states for their failed actions will not remedy the damages done, but it puts blame where it belongs and could act as a warning sign for future actions as did the victories against tobacco companies. Law could also be used to stop actions occurring right now, like the destruction of the Amazon rainforest, and should be further strengthened, for example by including ecocide in the Rome Statute as a crime against humanity.[39] In line with this, we should rethink the way we engage with the non-human world, for example via the recognition of Rights of Nature. Rights of Nature has come alive as part of constitutional transformations in Ecuador and Boliva, as part of settlements between colonialising Pākehā and Indigenous peoples of New Zealand, as well as in regional laws subjectifying waterbodies in the USA.[40] Such cultural and legal alteration in which nature is recognised to have rights is a way to move towards a post-extractivist world.[41]

Acting here and now

Judging from history we can recognise that current obstruction strategies bear the mark of previous, failed promises in consumer citizenship and market mechanisms. Strategies have aimed to displace action in time. To counter such imaginaries, we need to act here and now. That is, we need social movements, civil society and others to propose transformative policies that organisations, political parties and companies will realise.

Climate obstruction in its manifold materialisations necessitates the creation of an even stronger movement focusing on ending the use of fossil fuels.[42] Recent history, from women's rights to the anti-nuclear energy movement, illustrates that if we are to deal with both a heating planet and a transformation away from its causes, a similar social movement needs to be enacted. The global movement of Fridays For Future, which united many of the local, regional and national climate justice efforts, is one such example. Indeed, as illustrated by the aforementioned example of the anti-nuclear energy movement, such social movements have long been a major force regarding energy policy – with opposition to fossil fuel extraction and infrastructure now present in Europe (*Extinction Rebellion, Ende Gelände*) and around the world.[43] To insist on refraining from the use of a readily available energy source and to dismantle this industry is crucial as energy does not operate in the same way as other technologies: rather than replacing the old, new energy sources most often result in the addition of that source, without reduction in other sources.[44]

The nuclear issue is instructive in even more ways. The last time when humanity was faced by existential threat, non-proliferation was enacted. Today, such treaties are needed to immediately end exploration and expansion of new reserves.[45] Such a solution would be in line with recent research which has calculated how many new coal mines, how many new oil fields, and how many new gas holes can be opened if no carbon removal technologies are implemented: zero.[46] However, a non-proliferation fossil fuel treaty will only see the light of day through strong support by local, national and international social movements.[47]

Delay and a reliance on adaptive measures or new technologies have been ways to counter regulation since the rise of the environment as a global issue. To counter such secondary obstruction, the need to engage with existing technologies is crucial too. While the call to keep fossil fuels in the ground keeps us focused on the task at hand, the insistence on working with the means available is equally important. During the Covid-19 pandemic, we saw a few such promising signs. For example, due to social distancing, biking became more frequent which, in turn, led to roads being partially transformed into bike lanes in cities like Berlin, Budapest and Paris.[48] This shows that much can be done right now by simply privileging other types of transportation. Another important measure mentioned in the Working Group 3 report from IPCC is retrofitting houses to save energy. This was one of the major means via which Sweden cut its emissions after the 1970s. Today, we see similar calls in, for example, the Insulate Britain campaign. Redesigning infrastructure and limiting energy consumption in housing can draw on readily available technologies that lessen the dependency on, and the power of, the fossil fuel industry.

While it is true that we need technologies to limit the warming of the planet, many developers of carbon capture and storage technology, such as direct air capture, are fossil companies themselves.[49] Judging from historical actions, their goal is therefore most probably to enable the continuing burning of fossil fuels. For precisely this reason we are not convinced by the unproven carbon removal techno-solutions promoted as ways to reach the 1.5° or the 2°-degree targets.[50] This intention by fossil fuel companies to shape the technologies developed has, however, resulted in calls for control over this process and a return to more democratic forms of planning.[51] Suggestions in this vein have been present in the different forms of Green New Deals proposed and focused the attention of the climate movement towards engagement with wider parts of the population.[52]

This in turn shifts our attention to the importance of policies that take issues of justice into account. In other words, climate policy and action need to consider historical oppression and hierarchies and be created in a way that empowers and amplifies the low-impact practices that exist (which are often found outside of the modern Western industrial/breadwinner male norm).[53]

Changing cultures, engaging in society

When talking about everyday actions, the focus is often on changes in lifestyle and consumption affecting demand. Such suggestions can range from voluntary ones to regulatory policy. And indeed, within the field of degrowth studies, policies are a key theme aimed at limiting consumption.[54] One problem with this framing is that it plays into the idea of environmental action as a limitation or a restriction on one's life and thus regularly meets opposition or fails to gather political momentum among a majority of voters. One way to address this could be to frame environmental action in terms of co-benefits.[55] For example, while the expansion of public transport or the limitation of car traffic in inner-city areas would restrict the possibility for to drive a personal car, it would simultaneously reduce particle pollution and result in safer streets. Acting on climate change means changing destructive practices so widespread today, to less destructive ones and furthermore entails rethinking what a good life is. Indeed, changing identities, values and emotional entanglements away from an industrial modern petroculture.[56] Nevertheless, ideas regarding a feminist green new deal[57] and leadership training programmes moving away from, for example, industrial/breadwinner masculinities show promising results on issues such as reconnecting to nature, increased caring practices and raised preparedness to address the climate crisis.[58]

However, regardless of how willing people may be to change their lifestyles, such change is difficult to accomplish if societal norms stand in its way and when the infrastructure does not enable it. To address this, it has been argued that structural changes should come first, with changes in norms to follow as happened, for example, in the case of indoor smoking ban. However, this is not to say that laws should be designed and implemented without considering public opinion. To exemplify, carbon taxes tend to gain low support from the public and can spur protests – even though they are a cost-effective way of promoting sustainable behaviour.[59] Here, policymakers could consider how to develop and talk about compensatory measures and how to support those who have no realistic alternatives for the polluting action, even if this may not be enough if the public does not trust the policymakers. Trust is an important factor explaining policy support, and the correlation between climate beliefs and carbon tax, which varies across Europe depending on the level of trust in the nation.[60] Thus, transparency and clarity are necessary when communicating policy proposals.

One way of promoting commitment among the public, industries and policy makers is to find new ways and contexts to communicate climate change. The efforts to raise awareness and spark engagement are often focused on fixing the damages obstructionist frames have caused and to communicate the reality of the climate urgency. Yet, the sheer availability of factual knowledge is clearly not enough.[61] After all, also misinformation continues to flow and thus "inoculation" against such campaigns by "prebunking" and training individuals to detect and respond to their misleading arguments could be one solution.[62] Moreover, it is important that climate communication creates a sense of trust and efficacy regarding our capability to collaborate and address the crisis.[63]

A wide range of evidence-based guidelines exist on how to effectively communicate the need for climate action and to counter obstruction. These include the need to find common ground as a point of departure (instead of simply telling people "where they should be" or ridiculing "where they are", though without sacrificing authenticity), an emphasis on how climate change affects our everyday lives in the present (instead of simply pointing to an apocalyptic future), by pointing out how climate action might in fact improve our live(lihood)s (instead of being preoccupied with bans of this and that) and by letting people know how they can meaningfully contribute to create a better future (instead of limiting communication to fear appeals).[64] These are just a few examples of possible communication tactics to be used, either in broader information campaigns and media or in private discussions. Selecting the most suitable ones will necessarily depend on the context and the audience, and several such toolkits already exist. For example, Climate Outreach[65] has developed Britain Talks Climate that segments the population into groups, based on their core beliefs, with the aim to build narratives that would resonate with their values and everyday concerns.[66] All this harks back to the need to not simply convey "more (accurate)" information as what is in fact needed is creative communication which motivates meaningful action while not denying the grim reality of the climate crisis.[67]

The need for speed

This book has been about obstruction, more precisely, obstruction of actions that could have prevented the unfolding climate and ecological crisis. The question we (and many others before us) thus wanted to shed light on is how this lack of necessary action has come to be. Why we, to return to Secretary-General of the United Nations António Guterres, keep "sleepwalking to climate catastrophe"?

While the brunt of the responsibility lies with the polluting industries and their intentional efforts to keep extracting, the problem also lies with a society that has failed to imagine and enact a world beyond the destructive practices which already affect life on Earth. Indeed, industries and political actors have kept fuelling our lack of imagination and even ramp up their efforts to extract more as we write.[68] There is thus need for speed, need for swift and pervasive action, including to leave fossil fuels in the ground and to phase in new infrastructures which open new paths. At the same time, however, re*acting* to the climate and ecological crisis calls for a slowing down, for deceleration, for example in areas such as mobility and agriculture. That is, for example, as soils have come under increased (environmental) pressure, they require more time and less high-energy inputs to feed human and non-human populations.

As such, the climate crisis and climate obstruction point to much broader questions of how we want to live, of how humanity wants to relate to and situate itself within its non-human environment. Instead of dismissing the warnings by almost every relevant scientist on Earth (which we and others have reported) as scaremongering, acknowledging the planetary changes humans have induced should thus, also, give rise to such fundamental reflections. And while we contemplate, let us also change the world by leaving fossil fuels behind.

Notes

1 Oreskes, N. (2004) "The scientific consensus on climate change", *Science*, 306(5702), pp.1686–1686. https://doi.org/10.1126/science.1103618.
2 IPCC (2021) "Climate change 2021: The physical science basis", AR6 WGI, Summary for Policymakers, p.6. Available at: https://www.ipcc.ch/report/ar6/wg1/downloads/report/IPCC_AR6_WGI_Full_Report.pdf [Last accessed May 12, 2022].
3 For examples of research from different disciplines see Almiron, N. and Xifra, J. (eds) (2020) *Climate Change Denial and Public Relations: Strategic Communication and Interest Groups in Climate Inaction*. Routledge, p. 268; ·Moreno, J.A. and Thornton, G. (2022) "Climate action obstruction in the Spanish far right: the Vox's amendment to the Climate Change Law and its press representation", *Ámbitos. Revista Internacional de Comunicación*, 55, pp.25–40; McKie, R.E. (2021) "Obstruction, delay, and transnationalism: examining the online climate change counter-movement", *Energy Research & Social Science*, 80, 102217. https://doi.org/10.1016/j.erss.2021.102217; Carroll, W.K. (ed) (2021) *Regime of Obstruction: How Corporate Power Blocks Energy Democracy*. Edmonton: Athabasca University Press; Brulle, R.J., Hall, G., Loy, L. and Schell-Smith, K. (2021) "Obstructing action: foundation funding and US climate change counter-movement organizations", *Climatic Change*, 166(1), pp.1–7. https://doi.org/10.1007/s10584-021-03117-w.
4 Leiserowitz, A. *et al.* (2022) *Global Warming's Six Americas, September 2021*. Yale University and George Mason University. New Haven, CT: Yale Program on Climate Change Communication; Kácha, O., Jáchym, V. and Brick, C. (2022) "Four Europes: climate change beliefs and attitudes predict behavior and policy preferences using a latent class analysis on 23 countries", *Journal of Environmental Psychology*, 81, 101815. https://doi.org/10.1016/j.jenvp.2022.101815.
5 Povitkina, M., Jagers, S.C., Matti, S. and Martinsson, J. (2021) "Why are carbon taxes unfair? Disentangling public perceptions of fairness", *Global Environmental Change*, 70, 102356. https://doi.org/10.1016/j.gloenvcha.2021.102356; Fairbrother, M. (2022) "Public opinion about climate policies: a review and call for more studies of what people want", *PLOS Climate*, 1(5), e0000030. https://doi.org/10.1371/journal.pclm.0000030; Macdiarmid, J.I., Douglas, F. and Campbell, J. (2016) "Eating like there's no tomorrow: public awareness of the environmental impact of food and reluctance to eat less meat as part of a sustainable diet", *Appetite*, 96, pp.486–493. https://doi.org/10.1016/j.appet.2015.10.011.
6 Jylhä, K.M. and Hellmer, K. (2020) "Right-wing populism and climate change denial: the roles of exclusionary and anti-egalitarian preferences,

conservative ideology, and antiestablishment attitudes", *Analyses of Social Issues and Public Policy*, 20, pp.315–335. https://doi.org/10.1111/asap.12203; Milfont, T.L. *et al.* (2013) "Environmental consequences of the desire to dominate and be superior", *Personality and Social Psychology Bulletin*, 39, pp.1127–1138. https://doi.org/10.1177/0146167213490805; Stanley, S.K. and Wilson, M.S. (2019) "Meta-analysing the association between social dominance orientation, authoritarianism, and attitudes on the environment and climate change", *Journal of Environmental Psychology*, 61, pp.46–56. https://doi.org/10.1016/j.jenvp.2018.12.002.

7 We acknowledge the justice aspect of a transition away from fossil fuels limiting the pace and ability to transition in different parts of the world as well as the importance of hydrocarbons in almost all our modern products.

8 Althor, G., Watson, J.E.M. and Fuller, R.A. (2016) "Global mismatch between greenhouse gas emissions and the burden of climate change", *Scientific Reports*, 6, 20281. https://doi.org/10.1038/srep20281; Schlosberg, D. and Collins, L.B. (2014) "From environmental to climate justice: climate change and the discourse of environmental justice", *WIREs Climate Change*, 5, pp.359–374. https://doi.org/10.1002/wcc.275.

9 Schlosberg, D. (2013) "Theorising environmental justice: the expanding sphere of a discourse", *Environmental Politics*, 22, pp.37–55. https://doi.org/10.1080/09644016.2013.755387.

10 Ghosh, A. (2021) *The Nutmeg's Curse*. Chicago: The University of Chicago Press, p.120.

11 Sultana, F. (2022) "The unbearable heaviness of climate coloniality", *Political Geography*, 102638. https://doi.org/10.1016/j.polgeo.2022.102638.

12 Mckie, R. *et al.* (2021) "Is climate obstruction different in the global south? Observations and a preliminary research Agenda", *CSSN Briefing, CSSN Position Paper*, 4.

13 Bradley, K. (2009) "Planning for eco-friendly living in diverse societies", *Local Environment*, 14(4), p.359.

14 Machin, A. (2019) "Changing the story? The discourse of ecological modernisation in the European Union", *Environmental Politics*, 28(2), pp.208–227. https://doi.org/10.1080/09644016.2019.1549780.

15 Aronczyk, M. and Espinoza, M.I. (2022) *A Strategic Nature: Public Relations and the Politics of American Environmentalism*. Oxford: Oxford University Press, chapter 3.

16 See discussion about interpellation in Malm, A. and The Zetkin Collective (2021) *White Skin, Black Fuel: On the Danger of Fossil Fascism*. London: Verso, pp.322–335.

17 Though some have tried, see Dennis, K. and Urry, J. (2009) *After the Car*. Cambridge: Polity Press.

18 Wilson, S., Carlson, A. and Szeman, I. (2017) *Petrocultures: Oil, Politics, Culture*. Montreal: McGill-Queen's Press-MQUP.

19 Charbonnier, P. (2021) *Affluence and Freedom*. Cambridge: Polity Press.

20 Norgaard, K.M. (2011) *Living in Denial: Climate Change, Emotions, and Everyday Life*. Cambridge, MA: MIT Press.

21 Norgaard, K.M. (2011) *Living in Denial: Climate Change, Emotions, and Everyday Life*. Cambridge, MA: MIT Press.

22 Brulle, R.J. and Norgaard, K.M. (2019) "Avoiding cultural trauma: climate change and social inertia", *Environmental Politics*, 28(5), pp.886–908. https://doi.org/10.1080/09644016.2018.1562138.
23 Jost, J.T. and Hunyady, O. (2002) "The psychology of system justification and the palliative function of ideology", *European Review of Social Psychology*, 13, pp.111–153. https://doi.org/10.1080/10463280240000046.
24 Feygina, I., Jost, J.T. and Goldsmith, R.E. (2010) "System justification, the denial of global warming, and the possibility of 'system-sanctioned change'", *Personality and Social Psychology Bulleting*, 36, pp.326–338. https://doi.org/10.1177/0146167209351435; Hennes, E.P. *et al.* (2016) "Motivated recall in the service of the economic system: the case of anthropogenic climate change", *Journal of Experimental Psychology: General*, 145(69), pp.755–771. https://doi.org/10.1037/xge0000148.
25 Ekberg, K. and Hultman, M. (fast track 2021) "A question of utter importance, The early history of climate change and energy policy in Sweden 1974-1983", *Environment and History*; Mårald, E. and Nordlund, C. (2019) "Modern nature for a modern nation: An intellectual history of environmental dissonances in the Swedish welfare state. *Environment and History*.
26 Malm, A. and the Zetkin Collective (2021) *White Skin, Black Fuel.*
27 Rockström, J. *et al.* (2021) "Identifying a safe and just corridor for people and the planet", *Earth's Future*, 9(4), e2020EF001866. https://doi.org/10.1029/2020EF001866.
28 Hickel, J. *et al.* (2021) "Urgent need for post-growth climate mitigation scenarios", *Nature Energy*, 6(8), pp.766–768. https://doi.org/10.1038/s41560-021-00884-9; For a more comprehensive approach on pathways to mitigate climate change and limit carbon emissions see IPCC (2022) "AR6 climate change 2022 mitigation of climate change". Available at: https://report.ipcc.ch/ar6wg3/ [Last accessed May 13, 2022].
29 Zamponi, L. *et al.* (2022) "(Water) bottles and (street) barricades: the politicisation of lifestyle-centred action in youth climate strike participation", *Journal of Youth Studies*, pp.1–22. https://doi.org/10.1080/13676261.2022.2060730.
30 Newell, P., Paterson, M. and Craig, M. (2021) "The politics of green transformations: an introduction to the special section", *New Political Economy*, 26(6), pp.903–906. https://doi.org/10.1080/13563467.2020.1810215.
31 Hanieh, A. (2021) "Petrochemical empire. the geo-politics of fossil-fuelled production", *New Left Review*, 130; Huber, M.T. (2013) *Lifeblood: Oil, Freedom, and the Forces of Capital.* Minneapolis: University of Minnesota Press; Mitchell, T. (2011) *Carbon Democracy. Political Power in the Age of Oil.* London: Verso; van Asselt, H., Moerenhout, T. and Verkuijl, C. (2022) "Using the trade regime to phase out fossil fuel subsidies" in Jakob, M. (ed) *Handbook on Trade Policy and Climate Change.* Edward Elgar Publishing. https://doi.org/10.4337/9781839103247.
32 Smil, V. (2017) *Energy: A Beginner's Guide.* London: Oneworld Publications.
33 Coady, D., Parry, I., Sears, L. and Shang, B. (2017) "How large are global fossil fuel subsidies?", *World Development*, 91, pp.11–27. https://doi.org/10.1016/j.worlddev.2016.10.004.
34 van Asselt, H., Moerenhout, T. and Verkuijl, C. (2022) "Using the trade regime to phase out fossil fuel subsidies".

35 MacGregor, S. and Paterson, M. (2021) "Island kings: imperial masculinity and climate fragilities" in Pulé, P.M. and Hultman, M. (eds) *Men, Masculinities, and Earth*. Cham: Palgrave Macmillan, pp.153–168.

36 Bond, P. (2018) "Natural capital, carbon trading and climate sanctions" in Jacobsen, S.G. (ed) *Climate Justice and the Economy: Social Mobilization, Knowledge and the Political*. Abingdon, Oxon: Routledge.

37 Merner, D.L., Franta, B. and Frumhoff, P. (2022) "Identifying gaps in climate-litigation-relevant research: an assessment from interviews with legal scholars and practitioners", *CSSN Research Report 2022*, Policy briefing.

38 Farrell, J., McConnell, K. and Brulle, R. (2019) "Evidence-based strategies to combat scientific misinformation", *Nature Climate Change*, 9(3), pp.191–195. https://doi.org/10.1038/s41558-018-0368-6.

39 Walters, R. (2022) "Ecocide, climate criminals and the politics of bushfires", *The British Journal of Criminology*, azac018. https://doi.org/10.1093/bjc/azac018; Chiarini, G. (2022) "Ecocide: from the Vietnam War to International Criminal Jurisdiction?", *Procedural Issues In-Between Environmental Science, Climate Change, and Law, Cork Online Law Review*; Carter, P.D. and Woodworth, E. (2018) *Unprecedented Crime: Climate Change Denial and Game Changers for Survival*. SCB Distributors.

40 Fitz-Henry, E. (2022) "Multi-species justice: a view from the rights of nature movement", *Environmental Politics*, 31(2), pp.338–359. https://doi.org/10.1080/09644016.2021.1957615; O'Donnell, E., Poelina, A., Pelizzon, A. and Clark, C. (2020) "Stop burying the Lede: the essential role of indigenous law(s) in creating rights of nature", *Transnational Environmental Law*, 9(3), pp.403–427. https://doi.org/10.1017/S2047102520000242; Jones, E. (2021) "Posthuman international law and the rights of nature" in Grear, A., Boulot, E., Vargas-Roncancio, I.D. and Sterlin, J. (eds) *Posthuman Legalities*. Edward Elgar Publishing. https://doi.org/10.4337/jhre.2021.00.04.

41 Costa, A., Breblac, D. and Lupberger, S. (2016) "Post-growth and post-extractivism: two sides of the same cultural transformation", *Alternautas*, 3(1). https://doi.org/10.31273/alternautas.v3i1.1027.

42 Otto, I.M. *et al.* (2020) "Social tipping dynamics for stabilizing Earth's climate by 2050", *Proceedings of the National Academy of Sciences*, 117(5), pp.2354–2365. https://doi.org/10.1073/pnas.1900577117.

43 Temper, L. *et al.* (2020) "Movements shaping climate futures: a systematic mapping of protests against fossil fuel and low-carbon energy projects", *Environmental Research Letters*, 15(12), 123004. https://doi.org/10.1088/1748-9326/abc197; Etchart, L. (2017) "The role of indigenous peoples in combating climate change", *Palgrave Communications*, 3(1), 17085. https://doi.org/10.1057/palcomms.2017.85.

44 York, R. and Bell, S.E. (2019) "Energy transitions or additions?: why a transition from fossil fuels requires more than the growth of renewable energy", *Energy Research & Social Science*, 51, pp.40–43. https://doi.org/10.1016/j.erss.2019.01.008; Jerneck, M. (2021) "Achieving sustainable production through creative destruction: reflections on a multidisciplinary project" in Swain, R.B. and Sweet, S. (eds) *Sustainable Consumption and Production, Volume I*. Cham: Palgrave Macmillan, pp.107–123.

45 See https://fossilfueltreaty.org/ [Last accessed May 12, 2022].

46 Pellegrini, L. *et al.* (2021) "Institutional mechanisms to keep unburnable fossil fuel reserves in the soil", *Energy Policy*, 149, 112029. https://doi. org/10.1016/j.enpol.2020.112029; Newell, P. and Simms, A. (2020) "Towards a fossil fuel non-proliferation treaty", *Climate Policy*, 20(8), pp.1043–1054. https://doi.org/10.1080/14693062.2019.1636759; Welsby, D., Price, J., Pye, S. and Ekins, P. (2021) "Unextractable fossil fuels in a 1.5 C world", *Nature*, 597(7875), pp.230–234. https://doi.org/10.1038/ s41586-021-03821-8.

47 Hestres, L.E. and Hopke, J.E. (2020) "Fossil fuel divestment: theories of change, goals, and strategies of a growing climate movement", *Environmental Politics*, 29(3), pp.371–389. https://doi.org/10.1080/096440 16.2019.1632672; Campos, I. and Marín-González, E. (2020) "People in transitions: energy citizenship, prosumerism and social movements in Europe", *Energy Research & Social Science*, 69, 101718. https://doi. org/10.1016/j.erss.2020.101718; LeQuesne, T. (2019) "Petro-hegemony and the matrix of resistance: what can Standing Rock's Water Protectors teach us about organizing for climate justice in the United States?", *Environmental Sociology*, 5(2), pp.188–206. https://doi.org/10.1080/23251 042.2018.1541953.

48 Kraus, S. and Koch, N. (2021) "Provisional COVID-19 infrastructure induces large, rapid increases in cycling", *Proceedings of the National Academy of Sciences*, 118(15), e2024399118. https://doi.org/10.1073/ pnas.2024399118.

49 See for example Exxon: Global Thermostat (2022) "Global thermostat renews agreement with ExxonMobil", 5 May. Available at: https:// globalthermostat.com/2022/05/global-thermostat-renews-agreement- with-exxonmobil/; Preem: Preem (2019) "Här ska Preem fånga in koldiox- iden", 19 March. Available at: https://www.preem.se/foretag/kund-hos- preem/hallbart-foretagande/har-ska-koldioxiden-fangas-in/; Equinor (n.d.) "Carbon capture, utilisation and storage". Available at: https://www. equinor.com/energy/carbon-capture-utilisation-and-storage [Last accessed May 12, 2022].

50 Anderson, K. and Peters, G. (2016) "The trouble with negative emissions", *Science*, 354, pp.182–183. https://doi.org/10.1126/science.aah4567; Fleming, J.R. (2021) "Excuse us, while we fix the sky: WEIRD Supermen and climate intervention" in Pulé, P.M. and Hultman, M. (eds) *Men, Masculinities, and Earth*. Cham: Palgrave Macmillan, pp. 501–513. https:// doi.org/10.1007/978-3-030-54486-7_24; Rodriguez, E. *et al.* (2021) "Tensions in the energy transition: Swedish and Finnish company per- spectives on bioenergy with carbon capture and storage", *Journal of Cleaner Production*, 280, 124527. https://doi.org/10.1016/j.jclepro.2020. 124527; Hansson, A. (2012) "Colonizing the future: the case of CCS" in Markusson, N., Shackley, S. and Evar, B. (eds) *The Social Dynamics of Carbon Capture and Storage*. Abingdon, Oxon: Routledge, pp.98–114.

51 Mazzucato, M. (2011) "The entrepreneurial state", *Soundings*, 49(49), pp.131–142. https://doi.org/10.3898/136266211798411183; Buck, H.J. (2021) *Ending Fossil Fuels: Why Net Zero Is Not Enough*. Brooklyn, NY: Verso.

52 Ajl, M. (2021) *A People's Green New Deal*. London: Pluto Press; Pettifor, A. (2019) *The Case for the Green New Deal*. London: Verso; Aronoff, K.,

Battistoni, A., Cohen, D.A. and Riofrancos, T.N. (2019) *A Planet to Win: Why We Need a Green New Deal*. London: Verso; Pendergrass, D. and Vettese, T. (2022) *Half-Earth Socialism: A Plan to Save the Future from Extinction, Climate Change and Pandemics*. London: Verso; Huber, M.T. (2022) *Climate Change as Class War: Building Socialism on a Warming Planet*. London: Verso.

53 Paulson, S. (2017) "Degrowth: culture, power and change", *Journal of Political Ecology*, 24(1), pp.425–448. https://doi.org/10.2458/v24i1.20882; Barca, S. (2020) *Forces of Reproduction: Notes for a Counter-Hegemonic Anthropocene*. Cambridge: Cambridge University Press; Paulson, S., DeVore, J. and Hirsch, E. (2022) "Convivial conservation with nurturing masculinities in Brazil's Atlantic forest" in Adloff, F. and Caillé, A. (eds) *Convivial Futures: Views from a Post-Growth Tomorrow*. Bielefeld: Transcript Verlag.

54 Kallis, G. *et al.* (2018) "Research on degrowth", *Annual Review of Environment and Resources*, 43, pp.291–316. https://doi.org/10.1146/annurev-environ-102017-025941.

55 Bain, P.G. *et al.* (2016) "Co-benefits of addressing climate change can motivate action around the world", *Nature Climate Change*, 6(2), pp.154–157. https://doi.org/10.1038/nclimate2814.

56 Nelson, J. (2020) "Petro-masculinity and climate change denial among white, politically conservative American males", *International Journal of Applied Psychoanalytic Studies*, 17(4), pp.282–295. https://doi.org/10.1002/aps.1638; Daggett, C. (2018) "Petro-masculinity: fossil fuels and authoritarian desire", *Millennium*, 47(1), pp.25–44. https://doi.org/10.1177/0305829818775817; Pulé, P.M. and Hultman, M. (eds) (2021) *Men, Masculinities, and Earth*. Allen, I.K. (2022) "Heated attachments to coal: everyday industrial breadwinning petro-masculinity and domestic heating in the Silesian home" in Iwińska, K. and Bukowska, X. (eds) *Gender and Energy Transition*. Cham: Springer, pp.189–222. https://doi.org/10.1007/978-3-030-78416-4_11; Keller, J. (2021) "'This is oil country': mediated transnational girlhood, Greta Thunberg, and patriarchal petro-cultures", *Feminist Media Studies*, 21(4), pp.682–686. https://doi.org/10.1080/14680777.2021.1919729.

57 MacGregor, S., Arora-Jonsson, S. and Cohen, M. (2022) "Caring in a changing climate: centering care work in climate action", Oxfam Research Briefs. Available at: https://policy-practice.oxfam.org/resources/caring-in-a-changing-climate-centering-care-work-in-climate-action-621353/ [Last accessed May 13, 2022].

58 Hedenqvist, R., Pulé, P.M., Vetterfalk, V. and Hultman, M. (2021) "When gender equality and earth care meet: ecological masculinities in practice" in Magnusdottir, L.G. and Kronsell, A. (eds) *Gender, Intersectionality and Climate Institutions in Industrialised States*. Routledge, pp. 207–225.

59 Povitkina, M., Jagers, S.C., Matti, S. and Martinsson, J. (2021) "Why are carbon taxes unfair?"; Drews, S. and van der Bergh, J.C.J.M. (2015) "What explains public support for climate policies? A review of empirical and experimental studies", *Climate Policy*, 16(7), pp.855–876. https://doi.org/10.1080/14693062.2015.1058240.

60 Fairbrother, M. (2022) "Public opinion about climate policies"; Fairbrother, M., Johansson Sevä, I. and Kulin, J. (2019). "Political trust

and the relationship between climate change beliefs and support for fossil fuel taxes: evidence from a survey of 23 European countries", *Global Environmental Change*, 59, 1023003. https://doi.org/10.1016/j.gloenvcha. 2019.102003.

61 Suldovsky, B. (2017) "The information deficit model and climate change communication", *Oxford Research Encyclopedia of Climate Science*. https://doi.org/10.1093/acrefore/9780190228620.013.301.

62 Cook, J., Lewandowsky, S. and Ecker, U.K.H. (2017) "Neutralizing misinformation through inoculation: exposing misleading argumentation techniques reduces their influence", *PLoS ONE*, 12(5), e0175799. https://doi.org/10.1371/journal.pone.0175799; Farrell, J., McConnell, K. and Brulle, R. (2019) "Evidence-based strategies to combat scientific misinformation", *Nature Climate Change*, 9, pp.191–195. https://doi.org/10.1038/s41558-018-0368-6; van der Linden, S., Leiserowitz, A., Rosenthal, S. and Maibach, E. (2017) "Inoculating the public against misinformation about climate change", *Global Challenges*, 1(2), 1600008. https://doi.org/10.1002/gch2.201600008.

63 Feldman, L. and Hart, P.S. (2016) "Using political efficacy messages to increase climate activism: the mediating role of emotions", *Science Communication*, 38(1), pp.99–127. https://doi.org/10.1177/1075547015617941; Roser-Renouf, C., Maibach, E.W., Leiserowitz, A. and Zhao, X. (2014) "The genesis of climate change activism: from key beliefs to political action", *Climatic Change*, 125(808), pp.163–178. https://doi.org/10.1007/s10584-014-1173-5.

64 Boykoff, M. (2019) *Creative (Climate) Communications: Productive Pathways for Science, Policy and Society*. Cambridge: Cambridge University Press, pp.211–214. https://doi.org/10.1017/9781108164047. More specifically, see, for example, Feinberg, M. and Willer, R. (2013) "The moral roots of environmental attitudes", *Psychological Science*, 24(1), pp.56–62. https://doi.org/10.1177/0956797612449177; Whitmarsh, L. and Corner, A. (2017) "Tools for a new climate conversation: a mixed-methods study of language for public engagement across the political spectrum", *Global Environmental Change*, 42, pp.122–135. https://doi.org/10.1016/j.gloenvcha.2016.12.008; Wolsko, C., Ariceaga, H. and Seiden, J. (2016) "Red, white, and blue enough to be green: effects of moral framing on climate change attitudes and conservation behaviors", *Journal of Experimental Social Psychology*, 65, pp.7–19. https://doi.org/10.1016/j.jesp.2016.02.005; Hornsey, M.J. and Fielding, K.S. (2017) "Attitude roots and Jiu Jitsu persuasion: understanding and overcoming the motivated rejection of science", *American Psychologist*, 72(5), pp.459–473. https://doi.org/10.1037/a0040437.

65 Climate Outreach is a team of social scientists and communications specialists who work with international bodies, governments, charities, community organisations, academic institutions, businesses and media and provide evidence-based material and education on how to engage the public with climate change to create a social mandate for climate action. See https://climateoutreach.org/.

66 Similar efforts are seen in other countries too, for example the Swedish group Talking Climate that provides research-based civil society education material. Talking Climate is a non-profit organisation that train non-violent

interventions to the ecological crisis through workshops and trainings. See https://klimatprata.se/en/index.php.
67 Ojala, M. (2012b) "Hope and climate change: the importance of hope for environmental engagement among young people", *Environmental Education Research*, 18(5), pp.625–777. https://doi.org/10.1080/13504622.2011.63715 7; Ojala, M. (2012) "Regulating worry, promoting hope: how do children, adolescents, and young adults cope with climate change?", *International Journal of Environmental & Science Education*, 7(4), pp.537–561.
68 Kiln.it (2014) "Carbon map – Which countries are responsible for climate change?", *The Guardian*, 23 September. Available at: https://www.theguardian.com/environment/ng-interactive/2014/sep/23/carbon-map-which-countries-are-responsible-for-climate-change [Last accessed May 13, 2022].

Index

Printed in the United States
by Baker & Taylor Publisher Services